Hidden In Plain Sight 9

Andrew Thomas studied physics in the James Clerk Maxwell Building in Edinburgh University, and received his doctorate from Swansea University in 1992.

His *Hidden In Plain Sight* series of books are science bestsellers.

Also by Andrew Thomas:

Hidden In Plain Sight
*The simple link between relativity
and quantum mechanics*

Hidden In Plain Sight 2
The equation of the universe

Hidden In Plain Sight 3
The secret of time

Hidden In Plain Sight 4
The uncertain universe

Hidden In Plain Sight 5
Atom

Hidden In Plain Sight 6
Why three dimensions?

Hidden In Plain Sight 7
The fine-tuned universe

Hidden In Plain Sight 8
How to make an atomic bomb

Hidden In Plain Sight 10
How to program a quantum computer

Hidden In Plain Sight 11
The logic of consciousness

Hidden In Plain Sight 12
Consciousness and the steam engine

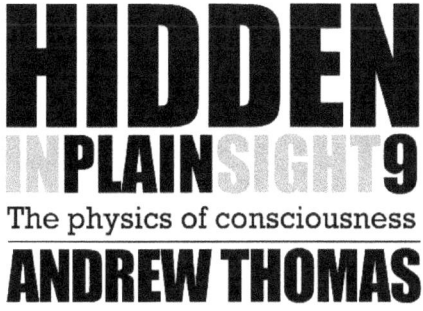

HIDDEN
IN PLAIN SIGHT 9
The physics of consciousness

ANDREW THOMAS

Hidden In Plain Sight 9

ISBN-13: 978-1984115072
ISBN-10: 1984115073

CONTENTS

"No problem can be solved from the same level of consciousness that created it."

- ALBERT EINSTEIN

1

THE PHYSICS OF CONSCIOUSNESS

Here are six questions. I would like you to answer either "yes" or "no" to each question:

1) Is a human being conscious? (Presumably you would answer "yes" to this question).

2) Is a chimpanzee conscious?

3) Is a dog conscious?

4) Is an ant conscious?

5) Is a computer conscious?

6) Is the internet conscious?

Maybe you answered "yes" to all six of those questions, but I imagine most people would have started saying "yes" then at some point they would have felt that a threshold had been crossed and they would have started to say "no", the object was not conscious.

I am not presenting these questions as a test to see if you give "right" or "wrong" answers. I certainly do not know the correct answers myself, and I do not even know if there is such a thing as a right or wrong answer to a few of these questions. I am merely presenting these questions to emphasize how difficult it is to define consciousness, and how difficult it is to detect the presence of consciousness.

Even though it is very difficult to define a feeling, we might ask "what is consciousness?" It might be said it is an awareness, a perception of one's self and the world around you. It is an awareness that you exist as an independent thinking entity.

I hope you get the message that it is very difficult to describe consciousness in anything but very vague terms. However, that is not to say the task of recognising consciousness is impossible. In the late 1960s, the psychologist Gordon Gallup was shaving himself in the mirror when he got the idea for a consciousness test which is now called the *mirror test*. The aim of the test is to determine if an animal has visual self-recognition, which is believed to be a measure of self-awareness. In the test, a mark is discreetly written on the forehead of an animal (for example, a chimpanzee) so that the animal is not normally capable of seeing the mark. The animal is then given access to a mirror and the reaction of the animal is observed. If the animal uses the mirror to recognise and investigate the mark, then that would appear to show that the animal perceives the reflected image as itself. This seems to suggest that there is some inner awareness of self – something which is often considered to be a key element of consciousness.

The following image shows an actual example of the mirror test being performed on a baboon:

However, so far only humans, chimpanzees, dolphins, and a single very clever elephant have passed the mirror test. Are we really suggesting that only those animals are conscious? Are we suggesting that dogs, cats, bears, and all those other animals are not conscious? Is the mirror test a reliable test of consciousness? In other words, is self-awareness a necessary factor of consciousness? We shall be considering these questions later in this book.

Perhaps it is easiest to admit that consciousness is best described in fuzzy terms as just a feeling, without strict definition: if you are currently wondering what consciousness is, then you must be conscious. It is the feeling you are having right now.

It is this vague and subjective definition of consciousness that has made it an unattractive and unpopular subject for physicists. And the few physicists who have dared to treat the subject seriously have generally, unfortunately, not found their work particularly well-received by the physics community. The real problem with the field is that it is not possible to test – or, more accurately, falsify – hypotheses about consciousness. Everyone seems to have their own personal hypothesis, their own personal theory. However, unlike physics, it is not possible to test those theories (how can you perform experiments on "feelings"?) to discover which hypotheses are incorrect. So, unlike physics, it is not possible to winnow-out the weak theories: there is no natural selection, no "survival-of-the-fittest" process for those theories. As a result, not only does the field not advance, but it gains a reputation for being unscientific – and best avoided by physicists. The field has therefore mainly been studied by neuroscientists and philosophers, with several valuable new insights coming from the neuroscientists. We shall also be seeing later that the neuroscientists are developing tests for consciousness, so consciousness research should no longer be considered "untestable".

This book, however, will take a slightly different approach. In this book, we will be tackling the problem from the viewpoint of physics, mathematics, computer science and artificial intelligence – and even some electronics.

But, before we start our quest to uncover the mystery of consciousness, there are a couple of principles (or problems) of which you should be aware ...

The mind-body problem

In the 17th century, René Descartes presented a rather direct explanation for consciousness. Descartes declared that the mind was made of fundamentally different "stuff" to the rest of the body. The mind was non-physical, and was distinctly separable from the rest of the body. This philosophy is called *dualism*.[1] Descartes even suggested a region in the brain where the mind might be found: the pineal gland, which is a small gland in the centre of the brain.

The motivation for Descartes philosophy was that the contents of our minds seem fundamentally different from the physical world we see around us. We might see a rock, and we know it is not conscious. We might see a more complex object, like a clock or a car, but we still know that those objects are not conscious. Somehow, the contents of our minds seem different. We do not seem tied to the physical world, the world of matter. Instead, our minds seem somehow immaterial, insulated from the material world, somehow floating in a higher realm of thought.

And in that higher realm, our insulated minds cannot be damaged by the physical world. As our bodies grow old and maybe painful, our minds can stay young, apparently impervious to the ravages of time. I have often heard an old

---

[1] In order to analyse the problem of consciousness, the approach of philosophers seems to be to call everything an "ism". As a result, we get *functionalism, computationalism, representationalism, materialism,* and many more "isms". Frankly, I consider it all to be excessive *classificationism.*

person say that they feel the same in their mind as they did when they were young – I feel the same way myself (though I do not yet consider myself old). It is as if the physical world cannot affect our mind.

But if that is really true – if the mind is immaterial and the physical world cannot affect the mind – then that raises a problem. The problem comes when we consider the other direction. Because if our minds really live in a higher-realm, unaffected by the physical world, then how on Earth can our minds ever affect the physical world? How can the immaterial consciousness in our heads move our arms and legs, for example?

This apparent paradox is called the *mind-body problem*. It is similar to the question of how ghosts in movies can walk through walls, but conveniently do not fall through the floor.

So, according to the mind-body problem, the challenge lies in the interface:

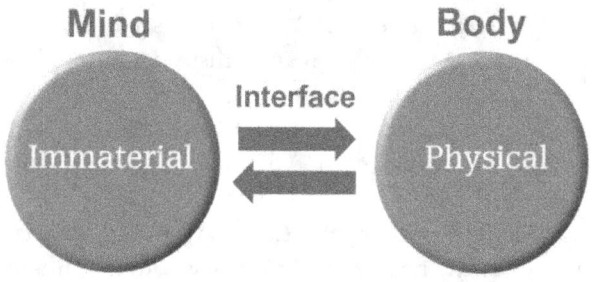

Substrate-independence

The second principle I would like to describe to you is based of the idea that there is nothing particularly special about biological cells: consciousness could still arise in some different underlying hardware. This idea seems eminently sensible: if you read my seventh book you will know that physics tells us that all electrons are exactly the same, and all protons are exactly the same. There is surely nothing special about the particles that form the atoms of biological cells.

This principle also seems to tie-in with our own experience. Our brains might be made of material with the consistency of rice pudding, but in our minds we do not feel as though we are made of "brain stuff" – we do not feel as though we are made of rice pudding. In fact, we feel completely independent of our brain hardware. We would surely feel the same even if our consciousness was transported to completely different hardware.

This principle – that consciousness is independent of the underlying hardware – is called *substrate-independence*. The principle of substrate-independence is described by the neuroscientist David Eagleman in his book *The Brain*:

> *If that turns out to be true, then in theory you could run the brain on any substrate. As long as the computations chug along in the right way, then all your thoughts, emotions, and complexities should arise as a product of the complex communications within the new material. In theory, you might swap cells for circuitry, or oxygen for electricity: the medium doesn't matter, provided that all the pieces and parts are connecting and interacting in the right way. In this way, we may be able to "run" a fully functioning simulation of you without a biological brain.*

A very similar standpoint is expressed by the neuroscientist Christof Koch in his book *Consciousness*:

> *Functionalism applied to consciousness means that any system whose internal structure is functionally equivalent to that of the human brain possesses the same mind. If every axon, synapse, and nerve cell in my brain were replaced with wires, transistors, and electronic circuitry performing **exactly** the same function, my mind would remain the same.*
>
> *It is not the nature of the stuff that the brain is made out of that matters for mind, it is rather the organization of that stuff – the way the parts of the system are hooked-up, their causal interactions. A fancier way of stating this is "Consciousness is substrate-independent".*

Taken to its logical conclusion, substrate-independence appears to suggest that there is nothing preventing a sufficiently-complex computer from becoming conscious. But is that really the case? We will be considering the question of whether computers can be conscious in Chapter Five.

The principle of substrate-independence seems to have gained widespread acceptance by consciousness researchers, and it would appear that any explanation of consciousness would likely need to possess the property of substrate-independence. That will be a guiding principle of this book.

So the mind-body problem and substrate-independence provide us with some guidance on our quest to explain consciousness. But where should we look for our solution? Surely, as physicists and scientists, there is only one place we should be looking …

Consciousness and the Standard Model

The discovery of the Higgs boson in 2012 completed the orthodox model of particle physics. That model is called the *Standard Model of particle physics*. The Standard Model contains a complete listing of the known elementary particles, together with a complete understanding of their interactions. Experiments have tested the predictions of the Standard Model with great accuracy, and there is no doubt that it is an accurate model of the subatomic world.

Here are the particles in the Standard Model:

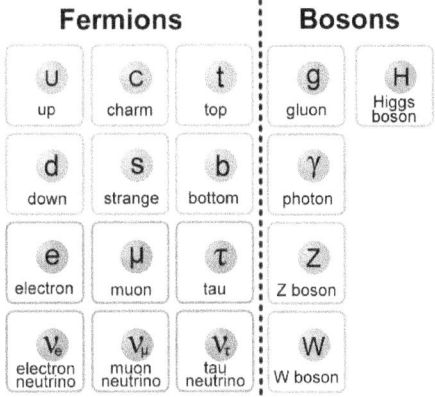

Because the Standard Model explains the universe with such accuracy, and as it has been tested so thoroughly, it makes it very difficult to propose any convincing new extension to the model – there really are very few gaps to be exploited. The most famous of these "Beyond the Standard Model" (BSM) theories is *supersymmetry*, which proposed an additional suite of particles to be added to the Standard Model, but experiments at the LHC have found no such particles.

The overwhelming success and acceptance of the Standard Model seems to raise challenges for the previously-described philosophy of dualism, in which consciousness is suggested to be held in some vague "immaterial" substance in the brain. As scientists, vaguely stating that an additional substance exists which is not described by the Standard Model is not sufficient: we have searched for additional particles using high-energy accelerators and we have found nothing more.

So our challenge appears to be to try to find where this peculiar "immaterial" mind substance can be found within the Standard Model.

As described earlier, the mind-body problem raises the question of how something immaterial could ever interact with the world of matter. How can consciousness ever hope to move arms and legs? What would the interface look like between matter and non-matter? Well, perhaps surprisingly, the Standard Model has the answer as to whether or not something immaterial can affect something material. The solution is that it is absolutely no problem: yes, something immaterial can definitely affect something material.

To understand why this is the case, we need to realise that all of the particles in the Standard Model can be divided into one of two classes. Referring back to the previous diagram, you will see that the first class of particles is the *fermions*, which are the matter particles such as the electron and the proton which can combine to make atoms. Atoms are made of *matter* – the "stuff" which makes the objects around us, the objects we can touch and feel. The second class of particles is the *bosons*, which might be considered the immaterial particles ("immaterial" means "not made of matter"). As an example, a beam of light is composed of photons which are bosons – and you cannot touch or feel an immaterial beam of light.

However, it is known that a photon (an immaterial particle) can deflect an electron (a matter particle) in an effect called *Compton scattering*. The following diagram shows a photon deflecting the path of an electron (this is showing a vertex from a Feynman diagram):

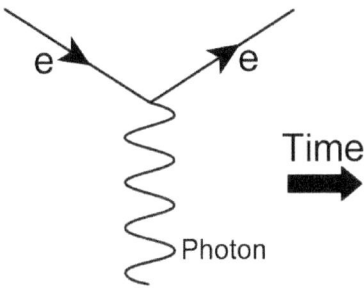

So here in one diagram we have an answer to one question which is often raised about the mind-body problem: "Could it ever be possible for something immaterial to affect something material?" The answer – according to the Standard Model – is yes it can, and this is how. The question raised earlier was how could there possibly be an interface between the material and the immaterial. Well, the previous vertex describes just such an interface.

The general point I am trying to make is that it **is** possible to cross seemingly unbridgeable gaps between the material and the immaterial, and we will be returning to this theme towards the end of this book.

However, elementary particles are not considered to have any internal structure and are surely not capable of sustaining consciousness as individual units. Also, attempting to tie consciousness to any particular form of hardware (e.g., a particular type of particle) appears to be at odds with our intended aim of finding a substrate-independent solution which would be independent of any particular underlying hardware.

So it would appear we have drawn a blank in our attempt to incorporate consciousness into the Standard Model. But we should not give up yet.

Because there is something else …

There is a different substance, a different candidate to be considered the "stuff" from which our consciousness is made.

This substance is completely tied to the physical world, it is therefore represented in the Standard Model, but it also appears to inhabit a realm above the physical world. What is more, it possesses the desired substrate-independence because it can traverse the world of **both** fermions **and** bosons – the material and the immaterial – giving an impression of how it might represent a potential solution to the mind-body problem.

This substance never decays, never grows old. In that respect, it is impervious to the damage inflicted by the physical world. But it is also rigorously defined both in physics and mathematics, and it now even has its own law of conservation – which is considered to be on the same level as the law of conservation of mass or energy.

But this substance seems grander than the purely physical: it can represent music and poetry, art and ideas. However, it can also start wars – and bring peace. We find this substance so fascinating that we can stare at it for hours every evening as it holds us spellbound.

In fact, this substance dominates our lives to such a extent that it has given its name to this entire current era of humanity.

So let me introduce you to this remarkable substance …

Information

463 West Street today is a rather inconspicuous twelve-story apartment block overlooking the Hudson River in Lower Manhattan. Everyday, people walk past staring into their smartphones, unaware that the technologies which makes their smartphones "smart" were developed in the building.

Until the 1960s, 463 West Street was the home of the Bell Telephone Laboratories, the largest industrial research centre in the United States. The technologies invented or developed there included talking movies, colour television, radar, and live video broadcasts. But the most monumental year of the Bell Telephone Laboratories came just after the war.

In 1948, there were two major developments in the Bell Telephone Laboratories which might initially appear unconnected, but acting together they were to create the modern world. The first announcement was of the invention of a tiny electronic device, described in the press release that "it may have far-reaching significance in electronics and electrical communication". That was definitely an understatement. The tiny device rather unimpressively resembled an insect with a small body and spindly legs. It was called the *transistor*.

Each of those people walking past the inconspicuous apartment block in Lower Manhattan holding an iPhone in their hand is actually holding two billion of those strange little transistor devices (in the Apple A11 microprocessor). The transistor was to revolutionise the world. We shall be considering the transistor in Chapter Three.

The second major discovery to emerge from the Bell Telephone Laboratories in 1948 was a scientific paper published in *The Bell System Technical Journal*. The paper had the title "A Mathematical Theory of Communication" and it was written by a thirty-two-year-old researcher named Claude Shannon. Shannon was thin, gaunt, with an intense gaze. They were the eyes of a man who could see the future:

If the transistor was to create the hardware for the iPhone, the idea in Shannon's paper was to supply the signal – and the content. The word "transistor" was a new term introduced by the Bell Telephone Laboratories in 1948. Shannon's paper introduced yet another new word: the *bit*.

At the time, the Bell Telephone Laboratories' revenue came from telephone sales, and their telephone network. The question was raised internally, "What do we sell? What do we carry?" The Post Office carried letters and parcels. What did Bell Telephones carry? Electricity along its wires? Surely not – it was not a power company. Shannon realised that what was being carried was conversations, but he needed some way to quantify the amount. Hence, Shannon devised the "bit" as a fundamental unit, just like the metre, or the litre, or the kilogram. But what did the "bit" measure?

Shannon said it himself. The bit was: "A unit for measuring **information**."

Up to that point, the word "information" had only been used very informally, with ambiguous meaning. Shannon redefined the term by giving it a very specific technical meaning: information was something you could measure in terms of the number of bits it took to describe a message. In his book titled *The Information*, James Gleick describes the impact of this redefinition of "information":

> And then, when it was made simple, distilled, counted in bits, information was found to be everywhere. It led to compact discs and fax machines, computers and cyberspace, Moore's law and all the world's Silicon Alleys.

The word "bit" is a contraction of "binary digit". A binary digit can have one of only two values: either 0 or 1. Bits can be strung together to make a larger binary number, which could potentially represent any form of information: a movie, words in a book, music. The more bits there are in a string, the more information is contained in that string:

```
101100110011000101101110101101011011101
```

When information is transmitted or stored, it is common to use an alphabet of symbols. The most widely-used alphabet is the Latin alphabet, with its modern day English, French, and Germanic variants. But the same principles apply to any alphabet of symbols.

Characters in the alphabet can be converted into bits in order to be transmitted or stored electronically. Let us now consider an example. Let us imagine we have a complete text document – written in English – which we want to transmit to a second computer in order for that second computer to display the document on its screen in its entirety. How many characters would we need in our alphabet?

We would obviously need the 26 letters of the English alphabet ("A", "B", "C", etc.) and the 26 lower-case equivalent letters ("a", "b", "c", etc.) requiring 52 characters in our alphabet. But there would also be other characters required for punctuation, such as the full stop (period), comma, and hyphen, and – of course – the "space" character between words. Other characters which might appear in a text document include the dollar sign, the ampersand, and bracket characters. So it appears our alphabet would require in the region of about a hundred characters, or maybe slightly more.

To cut a long story short, such an alphabet for representing text in computers has been constructed and it is called the ASCII code (American Standard Code for Information Interchange). There are 128 characters in the ASCII alphabet. Those characters are given numeric codes from (0 to 127) in order to be transmitted and recognised by the receiving computer.

In order to be transmitted as a stream of information, each character is converted into a seven-bit code. Why seven bits? That is because in the binary numbering system it is possible to represent a number from 0 to 127 using seven bits.[2] Here are a few of the ASCII characters together with their associated decimal codes and seven-bit binary codes:

---

[2] This is because $2^7=128$. I am sure there are many websites which can show you the basics of the binary numbering system if you are unsure. It is base-2 arithmetic which only uses the symbols 0 or 1.

ASCII Character	Decimal	Seven-bit binary
SPACE	32	0100000
$	36	0100100
&	38	0100110
A	65	1000001
B	66	1000010
a	97	1100001
b	98	1100010

So the total number of characters in an alphabet is given by 2^N, where N is the number of bits allocated to each character. In the previous ASCII example, the total number or characters would be 2^7, which is equal to 128.

But now let us consider the reverse question: if a single character from an alphabet is received, or a string of such characters, then how much information has been transmitted? The answer actually depends on the size of the alphabet. If a single character is transmitted from a very small alphabet (for example, a 0 being transmitted when the only possibilities are either 0 or 1) then the amount of information which is transmitted is less than if the character had come from a very large alphabet. This is because, in the latter case, you would be specifying a single symbol from a greater range of possibilities: more precise and more specific information is being transmitted. Shannon stated that the amount of information, H, transmitted by a series of n characters is given by:

$$H = n \log_2 W$$

where W is the size of the alphabet.

Hence, for the previous ASCII example, the amount of information contained in a single ASCII character (setting $n=1$ for a single character) would be $\log_2(128)$, which is equal to 7. The previous formula therefore tells us that the information transmitted by a single ASCII character is equal to seven bits of information.

By developing this equation to describe the amount of information, Shannon revealed how information could be defined and measured. Shannon's breakthrough is described by James Stone in his book *Information Theory*:

> *Before Shannon's paper, information had been viewed as a kind of poorly defined miasmic fluid. But after Shannon's paper, it became apparent that information is a well-defined and, above all, **measurable** quantity.*

And, if information could now be measured, that made it sound like a very real thing indeed. Information could now be processed as if it was a physical substance. It could be "compressed" like a gas (for example, to reduce the size of a movie file on a hard disk), or "filtered" like a liquid (for example, to extract interesting search results).

After Shannon, "information" resembled a physical substance.

After Shannon, "information" was a **thing**.

Maxwell's Demon

In our attempts to analyse consciousness, what we would really like to discover would be some way of representing conscious thought in a form which we could analyse scientifically. However, it would appear to be a hopelessly ambitious goal to be able to represent consciousness in that manner: how could you express a "feeling" in an equation? But while it seems an almost impossible task to quantify consciousness, it appears it **is** possible to quantify "thinking".

In this section, we are going to generate a scientific definition of "thinking". In doing so, we will gain an insight into precisely what activity the brain actually **does**.

The way we are going to define thinking is by considering a famous thought experiment which involves the following cute little gremlin:

In our thought experiment, the gremlin is going to be doing a lot of thinking. However, rather bizarrely, the gremlin is going to do **no actual physical work.** It might appear to be impossible for the gremlin to have any measurable effect on the physical world if he performs no actual physical work, but we shall see that the thought experiment describes a particular situation in which that is possible.

To analyse the thinking of the gremlin, we are not going to consider the details of the workings of the gremlin's mind. Instead, we are going to treat the gremlin's head as a "black box", and merely consider the external effects of the gremlin's thought processes. With no physical work being performed, we can be sure that all observed effects must be purely due to the intellectual efforts inside the gremlin's head. We are then going to define "thinking" as the process which must have occurred in the gremlin's head in order to produce the observed effects.

So let us consider the thought experiment.

In 1867, the great Scottish physicist James Clerk Maxwell wrote a letter entitled "On the Decrease of Entropy by Intelligent Beings". In the letter, Maxwell proposed a famous thought experiment called *Maxwell's Demon*.

The thought experiment is shown in the following diagram. You can see it involves our gremlin (the eponymous "demon" in Maxwell's experiment). But this is not a nasty demon – he is more like a cute gremlin (from the film of the same name) who cannot stop fiddling with things:

The diagram shows Maxwell's demon lurking over a chamber which is divided into two compartments. There is gas in both compartments, and initially the temperature of the gas is the same in both compartments. There is an airtight trapdoor in the wall separating the two compartments, and the demon is in charge of opening and closing the trapdoor.

The demon is selected because the demon is supposed to have beyond-human superpowers which allow him to see individual molecules in motion, and incredibly fast reflexes which allow him to open the trapdoor to let individual molecules pass through into the opposite compartment.

The molecules of the gas are all moving randomly, though some of the molecules are moving faster than other molecules. Remember that the demon has super-powers, and is capable of seeing individual molecules. As you can see in the previous diagram, when the demon sees a fast-moving molecule approaching the trap door from the right-hand compartment, the demon activates the trap door so as to let that faster-moving molecule pass to the left-hand side. Similarly, the demon lets slower-moving molecules pass from the left-hand compartment to the right-hand compartment.

As a result, the left-hand side contains more faster-moving molecules and will become warmer, while the right-hand side will become cooler. This result appears to be in conflict with the second law of thermodynamics which states that heat will always tend to move from a warmer region to a cooler region, thus eventually equalising the temperatures. Because of the intelligent action of the demon, you could even take advantage of this eventual separation of temperature by running a heat engine as the heat passes from the warmer compartment to the cooler compartment.

But there lies an apparent mystery. It appears we have discovered a source of perpetual power, a steam engine which could run forever. As James Gleick describes: "A demon who could catch the fast molecules and let the slow molecules pass would have a source of useful energy, continually refreshed."

The mystery becomes even deeper if we consider the trapdoor to be completely frictionless. In physics, "work" is defined as the distance an object is moved against a force. If the trapdoor is frictionless, then it will exert no force against the hand of the demon, and therefore the demon will have to perform no actual physical work in order to open and close the door. James Clerk Maxwell wondered how energy had been produced **"and yet no work has been done,** only the intelligence of a very observant and neat-fingered being has been employed."

As Maxwell states, no physical work has been done. The only effort has been the intellectual effort – the "intelligence", as Maxwell says – of the demon. Has the clever demon really reversed the laws of physics purely by thinking?

Well, to a certain extent, yes he has. It is certainly true that the gas in the left compartment will become warmer, and the gas in the right compartment will become cooler, in apparent violation of the second law of thermodynamics. To be more precise, the second law of thermodynamics states that the amount of *entropy* in a system will always increase, where we might interpret entropy as being the amount of disorder in a system. By ordering the gas according to temperature – warmer molecules in the left-hand side, cooler molecules in the right-hand side – the demon appears to have reduced the entropy of the system.

However, the second law of thermodynamics is not violated. This is because if we consider the system as a whole – including the demon – we find that the total entropy of the system has increased, in line with the second law of thermodynamics. The demon has to perform some work in his brain – even just by looking at the molecules and calculating their velocity. When the calculations are performed, it is realised that the increase in entropy inside the brain of the demon is sufficient to outweigh the reduction in entropy due to his actions. This result was described the American physicist Carl Eckart: **"Thinking generates entropy".**

So this provides us with a way to quantify "thinking". All we have to do is measure the change in temperature of the gas, and – with no physical work having been performed – we then know that that change must have been entirely caused by the demon's intellectual effort. Therefore, that value provides us with an objective method of the amount of "thinking".

So that is quite a result. We may not be able to measure consciousness using this method, but it does appear theoretically possible to measure "thinking". It's a good start!

Now let us remember what was said at the start of this section. It was stated that we were going to define "thinking" as the process inside the demon's head. So what does the demon do? The demon has to obtain information about the velocities of the particles (either visually or by some similar method), he then has to calculate the speed of the particle and decide whether it is slow or fast, and then finally come to a decision as to whether or not to open the trapdoor. All of this represents the capturing of information, and the processing of that information. So we can define thinking as *information processing*, and we can consider the brain to be an **information processing unit.**

The material of our thoughts

So might information be the "stuff" of our thoughts?

Information certainly seems to fit the bill. For a start, information possesses the substrate-independence which we would expect from consciousness. As long as the particular model of information processing is preserved, it would appear possible that the same consciousness could be created in a range of different hardware. Substrate-independence also means that information is not tied to any particular type of particle in the Standard Model. Information can be stored as fermions (as electrons stored in electronic memory, for example), or transmitted as photons (as electromagnetic radio signals, for example). Hence, information spans both the material and the immaterial.

Information is also resolutely physical. Information always requires a physical substrate. Information is rigidly defined both in physics and in mathematics. But, in many ways, information appears to lie in a realm above the physical, being completely substrate-independent.

This possibility is described by the *Information Philosopher*:[3]

The "stuff" of mind is pure information. Information is neither matter nor energy, though it needs matter for its embodiment and energy for its communication.

———————————————

[3] The Mind-Body Problem, *The Information Philosopher*, **http://tinyurl.com/informationphilosopher**

The way I am presenting the concept of information, as existing in a substrate-independent "higher realm" from the material world might make it appear as though it is a mystical substance, similar to the "immaterial" substance of consciousness proposed by Descartes. And, indeed, in its behaviour we see that information does possess many of the qualities exhibited by Descartes' mystical substance. But the point I want to make is that information is also rigorously defined scientifically: we can talk about it, we can analyse it, it unquestionably **exists**.

Information appears to be a perfect fit for the material of our thoughts, the substance out of which our consciousness is made. We will be returning to consider information later in this book when we will see how a network based on information can create our thoughts, our self-awareness, and our consciousness.

2

A QUICK TOUR OF THE BRAIN

The importance of the human brain was realised over two thousand years ago when the Greek physician Hippocrates (of the Hippocratic Oath fame) said: "From the brain, and from the brain only, arise our pleasures, joy, laughter and jests, as well as our sorrows, pains, griefs, and tears."

That seems like a fairly accurate description. The brain is our central information processing unit which appears to be the centre of our thoughts, the seat of our consciousness. However, the brain also has to perform other functions of which our conscious mind is never aware. Regular signals from the brain keep our heart beating, and keep our lungs breathing. The brain is also responsible for controlling the release of hormones into the bloodstream. If we had to consciously remember to "breathe, breathe, breathe, ..." then our conscious minds would be too occupied to ever think of anything more productive. Hence, these *autonomic* output signals are produced by unconscious regions of our brain.

Similarly, the mass of sensory information entering the brain from the nerves would overwhelm our conscious mind if we had to deal with each signal individually. This would include monitoring body temperature, and monitoring carbon dioxide levels in the blood (which can then be used for regulating heart rate and breathing). Therefore, the unconscious brain has the job of filtering the huge number of sensory signals coming into the brain, and only passing the really important signals to the conscious part of the brain.

The power of the unconscious mind is illustrated in the medical phenomenon which has come to be known as *blindsight*. Blindsight occurs in people with healthy eyes who are blind because they have experienced damage in the region of the brain which is responsible for conscious processing of visual information. Even though the subjects are unable to see, the experiment involves showing them various pictures and asking to guess what the subject of the picture is. It would appear to be a pointless exercise, but the experiment shows that the subjects guess more accurately than would be expected from chance alone. This reveals the extent of the involvement of the unconscious visual system.

The unconscious signals coming out of the brain ("breathe, breathe, …") pass out through the spinal cord, and, similarly, the unconscious signals coming in to the brain come in through the spinal cord. For that reason, the brain appears to be organised with its unconscious processing region positioned at the top of the spinal cord (and is therefore able to intercept signals), while the higher-level conscious region of the brain is positioned on top of that unconscious region. The conscious region of the brain is then able to receive the filtered information from the unconscious part of the brain.

Another benefit of this arrangement is that the unconscious brain can respond to critical reflex reactions – for example, quickly pulling your hand out of a fire – without ever having to pass those messages up to the higher-reasoning levels.

In his book, *The New Science of Consciousness*, Paul Nunez considers the situation in which a small child runs out in front of the car you are driving. Your unconscious mind would slam on the brakes of the car as a reflex action, and this can be performed in 150 milliseconds – which is much too fast for the conscious mind to reason about the situation (imagine the mental dialogue: "Should I start to brake?" – far too much time wasted). In fact, conscious awareness of the child does not emerge until a full half-a-second (500 milliseconds). Your conscious mind is not aware of any delay – it is as if your consciousness is always lagging half-a-second behind the reality of the situation.

The following diagram shows the brain roughly separated into two regions: the unconscious processing region positioned at the top of the spinal cord, and the conscious region positioned on top of the unconscious region:

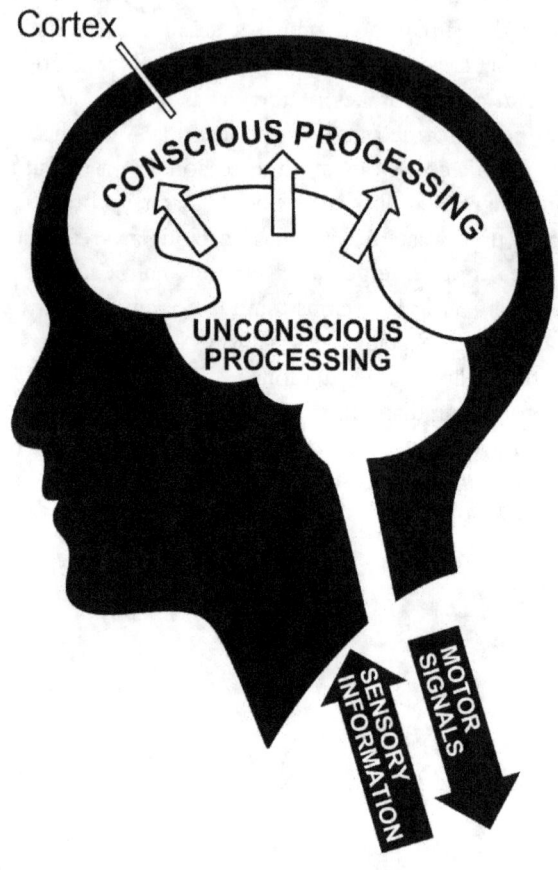

As can be seen on the diagram, the all-important region of the brain believed to be responsible for high-level thought processes is called the *cerebral cortex* (or simply the *cortex*). The cortex is a thin layer – only a few millimetres in depth – which is spread around the outside of the top region of the brain. It is the cortex that is the centre of our memory, awareness, thought, language, and consciousness.

To be more precise, the cortex is divided into two areas. Approximately 90% of the cortex is the *neocortex*, which is the region responsible for higher-level brain functions and is the region which interests us. The remaining 10% of the cortex is called the *allocortex* which is simpler and older in evolutionary terms ("neo" means "new"). In this book we will only be considering the neocortex. The terms "neocortex" and "cortex" are often used interchangeably. In this book, I will just be referring to the "cortex".

Because conscious processing appears to be confined to the cortex, we will only be considering the cortex in this book. In his book *On Intelligence*, Jeff Hawkins follows a similar strategy. In the following passage, Hawkins defends his preoccupation with the cortex, and is fairly dismissive of those who disagree:

> *I know I'm going to meet some resistance on this point, so let me take a minute to defend my approach before we get too far in. Every part of the brain has its own community of scientists who study it, and the suggestion that we can get to the bottom of intelligence by understanding just the neocortex is sure to raise a few howls of objection from communities of offended researchers. They will say things like: "You cannot possibly understand the neocortex without understanding brain region **blah**, because the two are highly interconnected like so, and you need brain region **blah** to do such and such."*

However, Hawkins stresses that it is the cortex which is the area which interests us from the point of view of consciousness:

> *We are going to focus most of our attention on the cortex. Almost everything we think of as intelligence – perception, language, imagination, mathematics, art, music, and planning – occurs here. Your cortex is reading this book.*

The cortex is composed of dense brain material: the *grey matter*. This material is made of brain cells called *neurons*. The neurons are so tightly-packed within the cortex that the thin cortex layer actually represents 40% of the total brain mass.

How can such a thin layer be responsible for all our high-level brain functions? Well, the cortex achieves this by expanding as far as possible in width and breadth, in other words, by increasing its surface area. So the cortex is a thin layer, but it spreads across the outer layer of the top of the brain.

We now see why the structure of the brain makes so much sense. The cortex needs to maximise its surface area, so it is positioned on the outer surface of the brain (the outer surface of a sphere has the greatest surface area). Conveniently, that means the cortex can cover the unconscious part of the brain, which, in turn, acts to filter the signals coming in from the centrally-positioned spinal cord.

So, as has been described, the cortex seeks to maximise its surface area. But the amount by which the cortex can expand is limited because it is surrounded by an hard, bony skull. The only way the cortex can grow in size is by pushing against the skull and wrinkling – thus giving the brain its characteristic wrinkled appearance. If you look at a human brain you only see about a third of the surface of the cortex because the majority is hidden in the wrinkles. If you could remove the cortex and spread it flat, it would be about the size of a small tablecloth, revealing how the wrinkling acts to increase the processing capacity of the brain. Many animals have perfectly smooth brains as their cortex does not grow enough to push against their skulls.

It seems that – when it comes to the brain – wrinkles really are a sign of wisdom.

Lobes of the brain

It is believed that different regions of the cortex are responsible for different brain functions. The division is not precise – there is some overlap – but the cortex is often divided into four *lobes*, each lobe having a distinct functional responsibility.

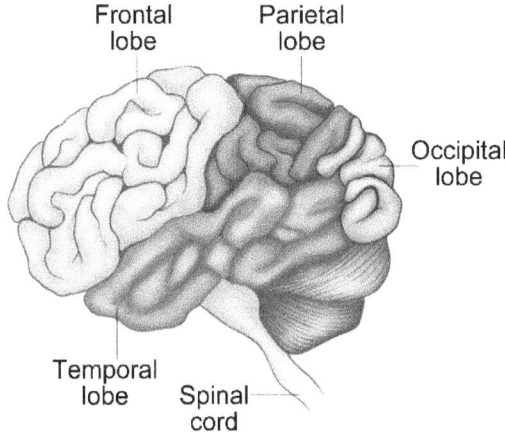

As an example of the function of the lobes, it is believed that the *frontal lobe* is where most of our imaginary thinking and problem-solving takes place, as well as language processing. It has been said that this is "the seat of our intelligence and the location of our personalities".[4] The frontal lobe is where we find most of the wrinkles in the cortex – the brain really does pack the processing power

---

[4] *The Secret World of the Brain*, Catherine Loveday.

where it is needed the most. As another example, the *occipital lobe* at the back of the brain is responsible for processing visual information – very much as if we have a cinema screen at the back of our head onto which the images are projected.

How have neuroscientists discovered which part of the brain is responsible for which type of behaviour? Well, originally it was discovered that people with brain injuries in different parts of the brain displayed behaviour which was characteristic of the site of the injury. The first example of this came from the work of the nineteenth century French physician Paul Broca. One of Broca's patients was nicknamed "Tan", because that was the only word he could say. In every other way, however, the patient appeared to be perfectly intellectually sound. When the patient eventually died in 1861, Broca conducted an autopsy and discovered that there was a lesion in his frontal lobe, and concluded that this area was responsible for language processing.

Nowadays, neuroscientists have more sophisticated methods of watching the brain in action. One method is Positron Emission Tomography (PET) scanning, in which the patient is injected with a mild radioactive material which is dissolved in glucose. The radioactive source decays via beta decay, emitting positrons (positrons are antimatter, and the principle behind PET scanning was considered in detail in my fifth book about particle physics). When the emitted positrons reach any normal tissue in the brain, they are annihilated (via matter/antimatter annihilation). This annihilation produces energy in the form of gamma rays. The gamma rays can be detected by the surrounding scanner, and the source of the radiation can be pinpointed.

While being scanned, the patient is requested to perform various tasks. To perform these varied tasks, different areas of the brain require more energy, and so use more of the injected glucose. Different regions of the brain then "light up" in the PET scan, thereby revealing which areas of the brain are responsible for the particular activity.

The neural correlates of consciousness

One of the biggest challenges in analysing consciousness is being able to detect the existence of consciousness in the first place. We might judge a person is conscious by whether or not they could respond to questions, or squeeze a hand when requested. However, this approach is not sufficient: some patients who are damaged in the lower brain but have no damage in the upper brain can be totally paralysed in a state known as "locked-in" syndrome. Even though they are conscious and aware of their surroundings, they are unable to physically respond.

So if consciousness cannot be detected via purely external responses, maybe it can be detected if we analyse the brain directly. With this end in mind, in the 1990s the American neuroscientist Christof Koch began working with Francis Crick – the molecular biologist who was the co-discoverer of the structure of the DNA molecule. Koch and Crick tried to identify the sectors of the brain which represented the active regions responsible for different conscious activities. Koch and Crick called these the *neural correlates of consciousness* (NCC), a term which seems to have become widely adopted.

According to Koch in his book *Consciousness*:

> *Imagine that you are looking at a red cube, mysteriously left in the desert sand, with a butterfly fluttering above it. Your mind apprehends the cube in a flash. It performs this feat because the brain activates specialized cortical neurons that represent color and combines them with neurons that encode the percept of depth, as well as neurons that encode the orientation of the various lines that make up the cube. The minimal set of such neurons that cause the conscious percept is the neural correlate of consciousness for perceiving this alien object.*

Koch is currently the Chief Scientific Officer at the Allen Institute for Brain Science in Seattle (funded by Microsoft co-founder Paul Allen). He is currently collaborating with Giulio Tononi of the University of Wisconsin in order to identify some of the neural correlates of consciousness. Their approach has involved EEG recording of the brain to determine which areas are active in conscious patients, and thereafter use that information to determine the presence of consciousness in other patients.

The situation is complicated by the fact that some areas of the brain are always active – in both conscious patients and unconscious patients. In order to take account of this fact, the brains of subjects who are in deep, dreamless sleep (i.e., in an unconscious state) were first analysed and then those areas (which would presumably be active all the time) were subtracted from images of brains in a conscious state. Logically, the resultant areas of the brain should be those responsible for consciousness.

Tononi and Koch are developing a theory known as *integrated information theory* (or IIT). The principle behind the theory is that the pattern of information in a conscious mind would show certain characteristics, and those characteristic patterns could be used to detect the presence of consciousness. Two key features of the pattern would be *integration* and *differentiation*. My understanding of IIT is that "integration" means that distant parts of the brain must be connected, and be able to talk to each other – otherwise the brain would contain many separate consciousnesses. So we might consider "integration" as referring to the large-scale patterns of activity in the brain. In contrast, "differentiation" refers to the small-scale patterns of activity. For example, we might be thinking of a visual scene containing many objects: cars, trees, houses, etc. So there is clearly a lot of detail, there. What is more, the representation of each of these small-scale objects must be clearly different from each other (otherwise we would be thinking of many objects which

36

were all the same). This requirement that the small-scale objects must be different is called *differentiation*. The following schematic diagrams show three possible arrangements of connectivity:

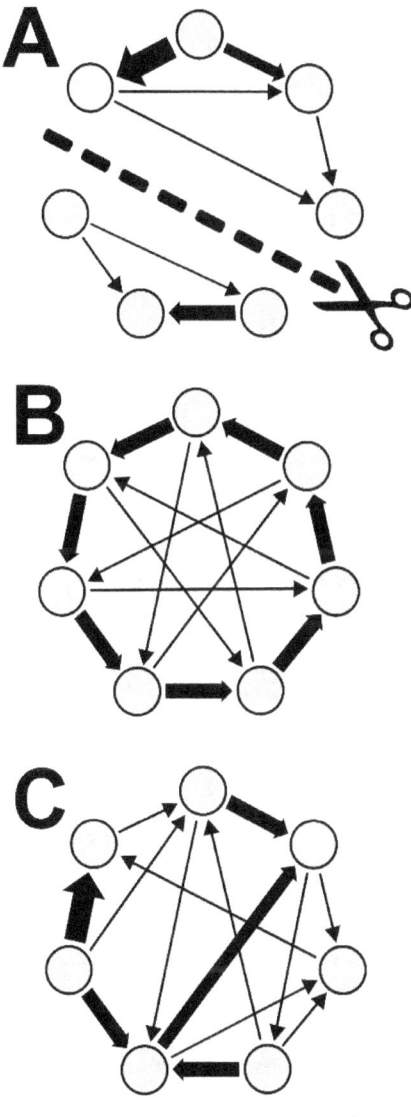

The top diagram (diagram A) shows a network with a lot of differentiation (with considerable variation in the types of connection) but a lack of integration (with two major regions of the network not being connected – you can see the network is cut by a pair of scissors). The middle diagram (diagram B) shows a great deal of integration (with all distant parts of the network being connected) but the perfect symmetry means there is a lack of differentiation. Finally, the bottom diagram (diagram C) shows a healthy balance between integration and differentiation. According to integrated information theory, in a healthy conscious mind, we would expect to find a balance of integration and differentiation in the EEG pattern, so this pattern could represent a conscious brain.

Giulio Tononi has developed a measure of integrated information which he has named Φ (the Greek letter "phi"). We shall be considering Φ in the next chapter on emergence.

In the November 2017 cover story of *Scientific American*, Christof Koch described the promise of IIT:

> *The development of several technologies in recent years has raised real prospects for detectors that meet the criteria for consciousness meters – devices useful in medical or research settings to determine whether a person is experiencing anything at all. This ability to detect consciousness could also help physicians and family members make decisions about how to care for tens of thousands of uncommunicative patients.*

In his book *Consciousness*, Koch also explains how this research will make consciousness research testable at last:

> *The bottom line is that these physiologic experiments are steadily narrowing the gap between the mind and the brain. Hypotheses can be put forth, tested and rejected or modified. And that is a great boon after millennia of sterile debate.*

Neurons

The human brain contains approximately 100 billion brain cells which are called *neurons* – which means that there are 14 times more neurons in your head than there are people on this planet. Neurons are the only cells which are able to transmit electrical signals. Neurons transmit those electrical signals to other neurons, forming a network (a network which eventually forms a brain).

At the heart of a neuron is a cell nucleus, just like any other cell. But the cell body is surrounded by small branch-like extensions which are called *dendrites*. These receive electrical signals from other cells. Therefore, these might be considered the "inputs" to the neuron. There is also one very long extension which comes out of the cell body which is called an *axon*. The axon can be extensively-branched. The axon proceeds to connect to other neurons, and it is a one-way connection. So the axon might be considered the "output" of the neuron.

The following diagram shows a neuron, showing the dendrites which surround the cell body, and the single axon output:

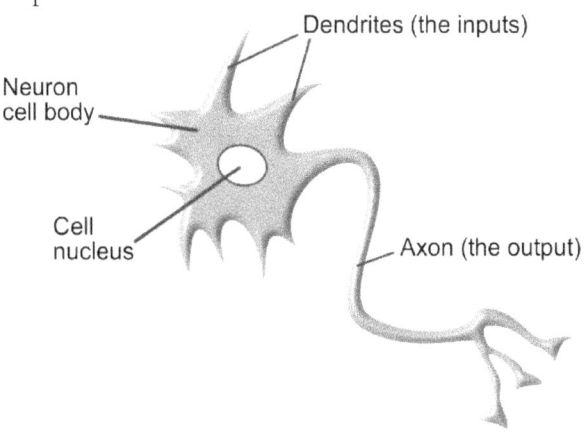

Dendrites (the inputs)

Neuron cell body

Cell nucleus

Axon (the output)

A crucial feature is the connection between two neurons, when the axon from a previous neuron connects with the dendrites of the next neuron. There is no direct physical contact between the two cells. Instead, there is a very small gap between the end of the previous neuron's axon and the next neuron's dendrites, and that gap is called a *synapse*. The synapse gap is just one forty-thousandth of a millimetre in width.

The following diagram shows two neurons which are connected together via synapses (shown in dashed circles):

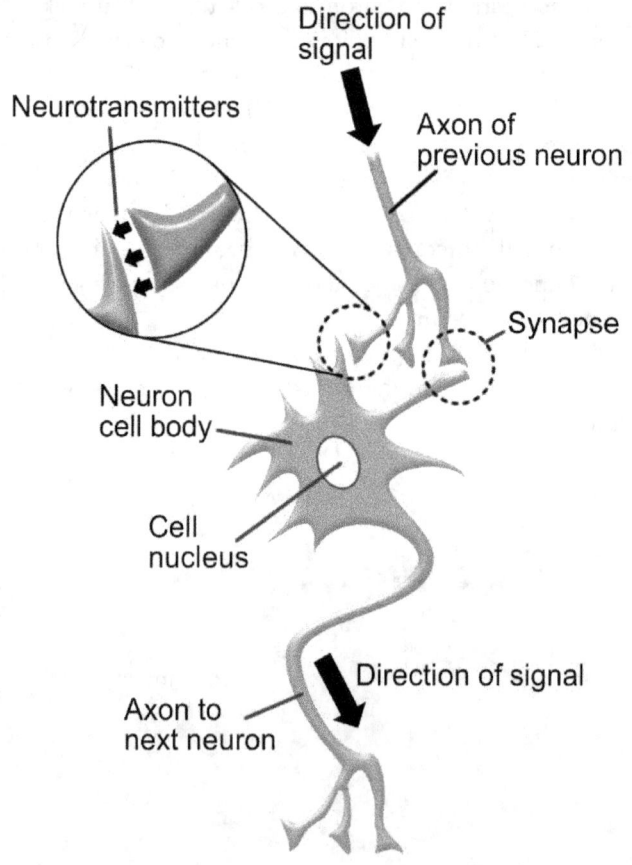

The synapse gap is filled with chemicals called *neurotransmitters*. Neurotransmitters play a vital role because they allow signals to bridge the gap of the synapse, thus allowing signals to be transmitted between neurons. It has been shown that these neurotransmitters play a vital role in our moods and our thinking processes. For example, it is known that variations in the amount of certain neurotransmitters (such as dopamine and serotonin) can cause mental health issues such as anxiety and depression.

Mind-altering drugs also target the neurotransmitters in the synapses. Stimulants, which include caffeine, nicotine, amphetamine, and cocaine, work by increasing the levels of the neurotransmitters dopamine and noradrenaline. Conversely, sedatives such as alcohol and Valium operate by reducing neurotransmitter activity in the synapse.

It might, therefore, be said that the construction of brain neurons with the synapses outside of the cell is an ingenious design feature. This makes the all-important synapses easily-accessible to the global release of chemicals into the general intercellular fluid of the brain. By releasing large amounts of neurotransmitters into the brain, it is possible for the body to trigger virtually instantaneous changes in mood and performance. For example, the release of noradrenaline increases the body's "fight-or-flight" response. And the release of natural opiates can lessen pain, which could be a life-saving strategy if the subject has to quickly run (on a broken leg) to escape a dangerous situation.

Inside the neuron

We have a remarkably full and accurate understanding of the internal structure of a neuron. What follows next is a description of the mechanism by which a neuron "fires" and transmits the output signal down its axon.

We know that messages are sent as electrical signals. It might be imagined that these messages would be formed of electrons passing down a conductive substance (like electricity passes down a wire). However, this is not the case. We will now see that a completely different – and apparently unnecessarily uncomplicated – method is used for sending signals down a neuron's axon to the next neuron. Indeed, in his book *The Emperor's New Mind*, the physicist Roger Penrose expresses his surprise that such a complicated method is used instead of the apparently obvious method of simply sending electrons down a "wire":

> *What form do the signals take as they propagate along nerve fibres and across the synaptic clefts? What causes the next neuron to emit a signal? To an outsider like myself, the procedures that Nature has actually adopted seem extraordinary – and utterly fascinating! One might have thought that the signals would just be like electric currents travelling along wires, but it is much more complicated than that.*

Instead, the signals are passed in the form of *ions*. An ion is an atom which has either positive electric charge (because it has lost an electron) or negative electric charge (because it has gained an electron). Signals are passed by the physical movement of these ionised atoms.

Common examples of ions found in neurons are sodium ions and potassium ions. Sodium has the chemical symbol

Na, while potassium has the chemical symbol K. Sodium and potassium are both metals, which means they have a single electron in their outer shells (it is that electron which allows electric current to pass through a metal). By losing that outer electron, a sodium atom can form a positive ion (Na^+), and a potassium atom can form a different positive ion (K^+). Other ions can be formed which have negative charge.

The body of each neuron is surrounded by a cell wall called a *membrane*. The membrane is punctured by many microscopic gates called *ion channels* which only allow certain ions to pass through them. These gates are opened or closed depending on the electric voltage (or *potential*) across them.

Because there are generally more negative ions inside the neuron than outside the neuron, there is a voltage difference of approximately -70 millivolts across the membrane, with the interior of the neuron being slightly electrically negative compared to the exterior environment of the neuron. This voltage is called the *resting potential*. This electric potential is enough to hold closed all of the ion channel gates.

However, when the neuron receives a large number of input signals on its dendrites, this can increase the amount of positive charge inside the body of the neuron. If this increase in positive charge increases beyond a certain threshold value, then it can overcome the voltage difference which was causing the sodium ion gate to remain closed (remember: it was the -70 millivolt voltage difference which was holding the gate closed). So when the threshold voltage is passed, the gate opens and the exterior sodium ions flood through the membrane into the neuron interior, attracted by the negative charge inside the neuron, as shown in the following diagram:

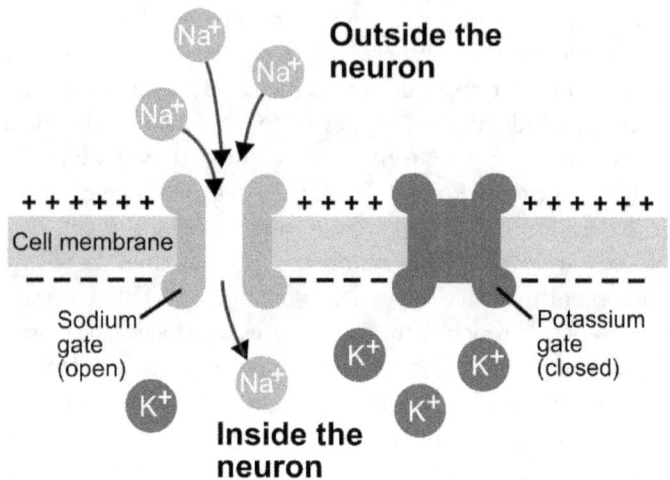

The influx of sodium atoms makes the interior of the neuron even more positively charged. The forms a positive feedback effect, and more sodium gates open. The effect is explosive, the opening of gates forming a ripple effect which passes down the neuron's axon at a speed of up to 100 metres per second. This fast moving spike of voltage represents the "firing" of the neuron and is called the *action potential.*

As the voltage continues to climb, a point is reached at which the sodium gate closes, and the potassium gates open. There is then an outflow of the positively-charged potassium ions, repelled by the positive charge inside the neuron, as shown in the following diagram:

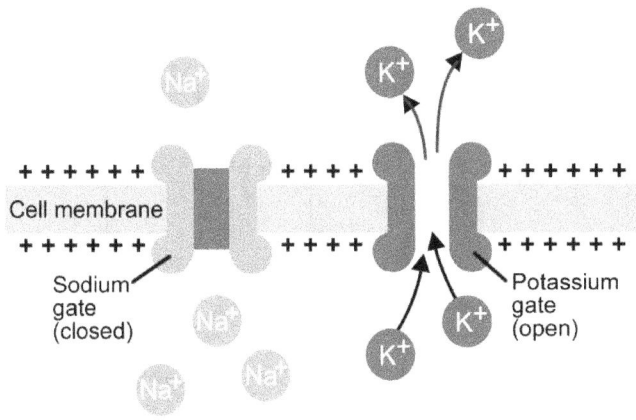

The effect of this outflow of positive charge is to restore the potential of the membrane back to its resting potential, ready for the next firing of the neuron.

This mechanism is not just used by brain neurons – it is also used by nerves to tell muscles when to fire. Whenever I come home after running, I always have an *electrolyte* drink. Electrolytes are ions including sodium, potassium, and magnesium. After exercise, you can lose salt through perspiration. Salt (sodium chloride) includes the electrolytes sodium and chloride. It is therefore important to top-up your body's store of electrolytes to avoid muscle cramp, in which the ability to send correct firing signals to the muscles is lost.

So why is such a complicated chemical process used to send messages down an axon, instead of simply sending an electrical signal down a "wire"? In the next chapter we will see a reason why this is possibly the case, and it is all to do with the -70 millivolts threshold voltage difference across the membrane. We will be seeing how this threshold voltage can result in "nonlinear" behaviour which would surely be necessary in any thinking machine.

In the last two sections, the two main methods by which neurons process information have been described. As Christof Koch said: "Two operations underlie information processing by neurons: the chemical transmission of information from one neuron to another at the synapses, and the generation of action potentials."

It is clear that the method by which signals are passed down the axon of a neuron is surprisingly complicated, with the arrangement of "gates" and "pumps" resembling nanotechnology. But it is also clear that we have a very complete and detailed understanding of these processes. We really do understand the workings of a single neuron remarkably well.

But that is not the challenge for us.

It is undoubtedly the case that consciousness is not held in any individual neuron. Individual neurons frequently die without obviously affecting our thought processes: we do not suddenly perceive any "dead pixels" in our mental images. Instead, consciousness arises from the interaction of billions of neurons: there may be 100 billion neurons in your brain, but each neuron is connected to as many as 10,000 other neurons. If you do the arithmetic you find that gives the extraordinary total of 1,000 trillion synaptic connections in the brain.

To suggest that we can understand consciousness by considering a single neuron – in however much detail – is to take a "reductionist" approach. Instead, there is a general consensus that consciousness arises as an "emergent" effect from the huge connectivity in the brain. Our challenge is to understand how consciousness emerges from this network.

In Chapter Four, we will be examining the severe challenges which emergent behaviours pose for our current way of doing science.

3

ARTIFICIAL NEURONS AND ELECTRONIC BRAINS

In this chapter, we will discover that neurons possess a very particular mathematical property which allows them to perform calculations, and to allow thinking to occur. It is undoubtedly a property which the components of all conscious systems must possess. We will also be seeing that it is a property which is possessed by electronic transistors, a property which allows those transistors to make computers – which are, after all, "thinking machines".

The property is called *nonlinearity*.

To understand nonlinearity, and why it is so important, let us first consider the related concept of *linearity*.

As stated in the previous section, a neuron can be considered as having a series of inputs (its dendrites) and a single output (its axon). To simplify this picture, let us draw a neuron as a simple block diagram:

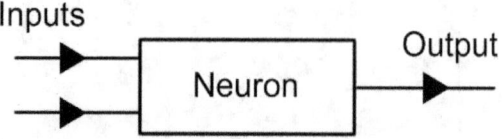

We might then ask how the output value varies as the input values are varied. We would probably imagine that as the input values increase, then the output value would increase proportionately. We might imagine the neuron simply summing all the input values and passing that total value to the output, or maybe multiplying that output by a constant number. In either of those cases, the output will be directly proportional to the input. As an example, the following graph shows this behaviour when the output value is equal to twice the total input value:

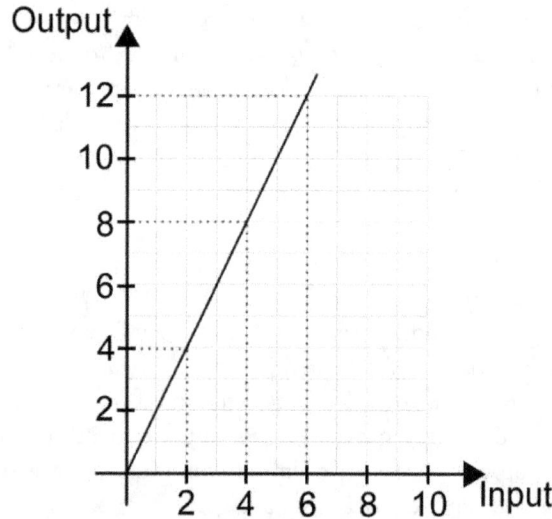

You can see in that case that the relationship between the neuron's input and output is given by the steep diagonal line.

A few values for input and output have been plotted on the graph. For example, when the input is equal to two, you will see that the output is four (2×2). And when the input is six you can see that the output is twelve (2×6).

Because the relationship between the neuron's input and output forms a perfectly straight line, we say that represents a *linear* relationship ("linear" meaning "like a line"). We might then call it a linear neuron.

In physics, many systems can be modelled as linear systems. For example, if you double the weight attached to a linear spring, then the spring would extend to twice its original length. Linear systems are easy to analyse mathematically, so physicists find it very convenient to work with linear systems. The reason that linear systems are easy to work with is because of the *superposition principle*. A description of this principle is presented on the Wikipedia page for the superposition principle:[5]

If input A produces response X and input B produces response Y then input (A + B) produces response (X + Y).

We can see that this principle is true on the previous graph. We can see that an input of 2 produces an output of 4, and an input of 4 produces an output of 8, so – according to the superposition principle – an input of 2+4 (which is 6) should produce an output of 4+8 (which is 12). And we can see that this is the case from the graph because an input of 6 does indeed produce an output of 12. So the superposition principle is indeed true for linear systems.

––––––––––––––––––––

[5] **http://tinyurl.com/superpositionprinciple**

The key point I am going to be making about linearity is that the superposition principle reveals that a brain could never be made out of these linear neurons. We will now see why that is the case.

The superposition principle reveals that it is possible to greatly simply any system which is made out of linear components. For example, it shows that if we had a system made out of two of these linear neurons, we could simplify it to a system of just one linear neuron.

The following diagram shows how this simplification can be achieved. It shows two linear neurons, which both act to double the value of the input in a linear manner (as described earlier). You can see that, when inputs of 2 and 4 are applied to the two neurons, the combined output is equal to 12.

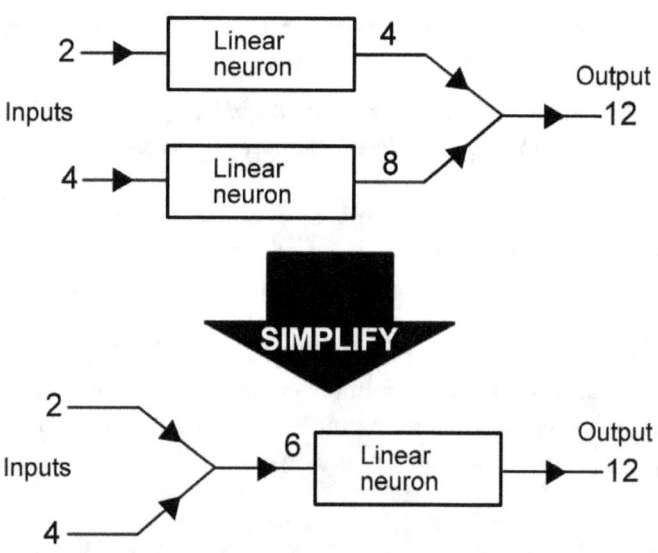

However, the superposition principle states that we could achieve the same result by using just one of those linear neurons, and summing the input – as you can see in the lower section of the previous diagram. You can see that with precisely the same inputs, the single neuron gives the same output as the output of the two neurons.

So there is no need to have two linear neurons – just one linear neuron would achieve exactly the same result. In fact, the superposition principle reveals that no matter how many linear neurons we have, in whatever configuration, the result can always be simplified down to a single neuron.[6]

To see this principle in action, let us consider a linear component from the field of electronics (we will be seeing that there are remarkable parallels between electronic components and neurons – in the same way that there are remarkable similarities between a computer and a brain). The electronic component we will be considering is a *resistor*. The resistor is a simple component whose behaviour might be thought of as "resisting" (i.e., reducing) the electrical current flowing through a wire. Here is a photograph of a typical resistor, with the resistance value (measured in ohms) being denoted by the coloured bands painted around it:

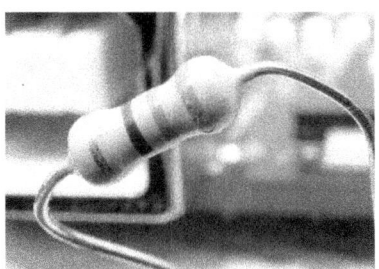

---

[6] Imagine a system made of four linear components. Two of the components could be simplified to just one component, and the other two components could also be simplified to one component. Which would leave you with two linear components. Those two components could then be simplified to just one component. So a system with any number of linear components can be simplified to just one component.

The symbol for a resistor in a schematic circuit diagram is a small rectangle, as shown in the following diagram:

The current through a resistor is proportional to the voltage applied across the ends of the resistor. The relationship is directly linear, which means if your drew a graph relating current to voltage, then it would be a perfectly straight line. This linear behaviour of the resistor is important because it means that the previously-described superposition principle applies if we have a network built of many resistors. That principle can be used to greatly simplify the network.

As an example, consider the following circuit diagram which contains seven resistors (and a battery on the left). The resistors have different values written next to them, measured in ohms. You will see that the symbol for the ohm is the Greek letter omega, Ω:

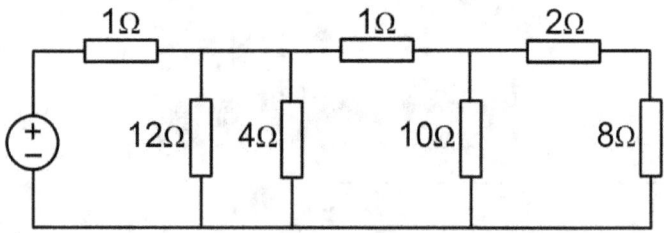

The superposition principle suggests that we could simply this network down to a single component – a single resistor. And that is precisely what we find. There are simple arithmetic formulas which are well-known to any electrical engineer which can reveal that the previous network of seven resistors is equal to a single resistance value of precisely 3 ohms:[7]

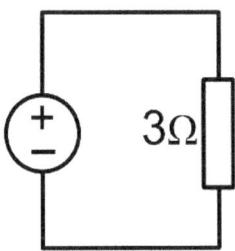

So a network consisting purely of a number of resistors can always be simplified to a single resistor – as revealed by the superposition principle.

If you were actually designing an electronic circuit, you could never create an interesting complex system with just one resistor: a single resistor can do nothing useful. Therefore, logically, you could never create a complex system with any greater number of resistors – as the situation would be equivalent to the single resistor case.

And the same principle applies to linear neurons: you could never create complex system with any number of linear neurons. Crucially, you could not build a brain with a single linear neuron, so, logically, **you could not build a brain with 100 billion linear neurons** (as the 100-billion-neuron brain could always be simplified to a single-neuron brain). For this reason, linear neurons could not be used to build a brain.

---

[7] If you are interested in how this simplification can be achieved, see the original example on the Khan Academy website: **http://tinyurl.com/simplifyingresistance**

So, if linear components on their own cannot be used to create complex systems, let us now move on to consider *nonlinear* components. As an example, the following graph shows the relationship between input and output when the output is the square of the input. You will see that the relationship between input and output is no longer described by a straight line – instead, it is a curved line. Hence, it is a nonlinear relationship:

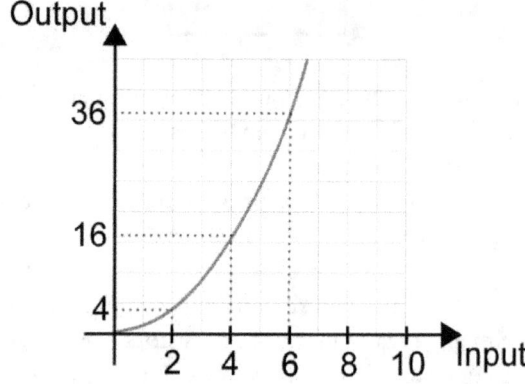

Let us see if the superposition principle still applies in this nonlinear case. Considering the previous graph, an input of 2 produces an output of 4, and an input of 4 produces an output of 16. If the superposition principle applies then an input of 2+4 (which is 6) should produce an output of 4+16 (which is 20). But, from the graph, we can see that is not the case: an input of 6 produces an output of 36 – not 20. So, crucially, **the superposition principle does not apply to nonlinear components.**

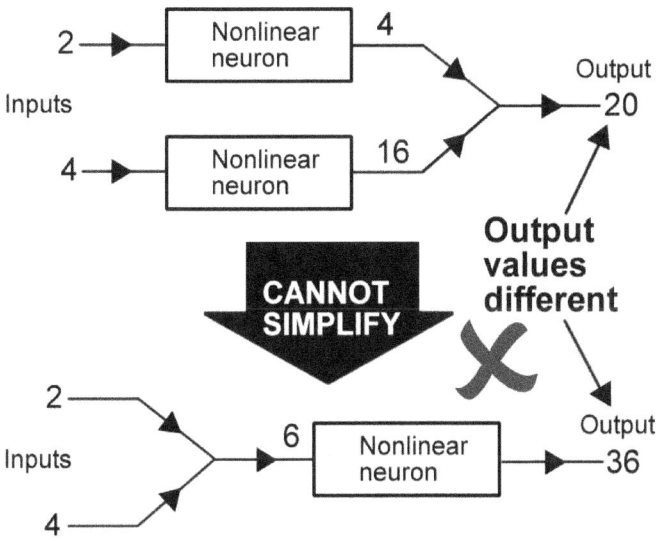

And, as you can see from the previous diagram, if the superposition principle does not apply then a network built from nonlinear components cannot be simplified (i.e., unlike linear components). It might sound like a bad thing to be unable to simplify a network, but if we want to create a large and complex network, it is very good news indeed. A network which cannot be simplified down to a single component is a network which can do very interesting things indeed.

Let us now consider an electronic component which is much more interesting than a resistor – precisely because it is nonlinear. This is a photograph of a *transistor* (its actual size would be only about two centimetres long in total):

It was described in Chapter One how the transistor was invented in the Bell Telephone Laboratories in 1948. It was invented by John Bardeen, Walter Brattain, and William Shockley, the three receiving the 1956 Nobel Prize in Physics for their invention. It could be argued that the transistor was the most important invention of the 20th century.

If we examine the appearance of the transistor in the previous photograph we see a small body and long, thin, flexible legs. Apart from the obvious difference in size, this is reminiscent of a brain cell – a neuron – described in the previous chapter. A neuron has a small cell body, a series of inputs (dendrites) and a long axon output. The similarity is not coincidental as the two different devices have to perform similar functions: receiving inputs from other similar units, and passing outputs to other similar units, forming a network of a vast number of components. In this chapter we will be seeing that the similarities between a transistor and a neuron are much more than just skin deep.

The material inside a transistor is called a *semiconductor*, and is often silicon or germanium. A semiconductor can operate as either an electrical conductor or an electrical insulator. In a transistor, the behaviour of the semiconductor depends on the voltage applied to one particular leg of the transistor, called the *base*. If a sufficiently high voltage is applied to the base, a current can pass between the other two legs. However, if no voltage is applied to the base, the transistor acts like an insulator and no current can pass between the other two legs.

The following diagram shows the schematic symbol for a transistor as used in circuit diagrams. The three legs are shown. The base leg is indicated, and the current flowing between the other two legs is also shown:

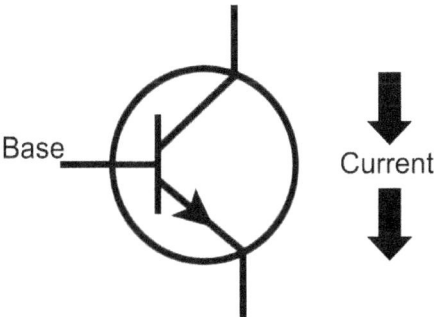

We will be seeing this diagram again later in this book.

Crucially, though, the behaviour of the transistor is nonlinear. Until the base voltage reaches 0.7 volts, no current can pass through the transistor. But once the 0.7 voltage is reached, the transistor "turns on" and current can pass.

This nonlinear behaviour is shown by the thick black line in the following diagram:

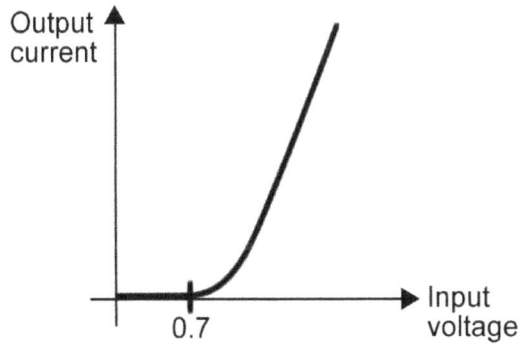

So the 0.7 volts acts as a threshold voltage. If you scan your memory, you might remember where we encountered a "threshold voltage" earlier in this book. We encountered a threshold voltage in the discussion of neurons at the end of the previous chapter. There it was explained how, when the neuron receives a large number of input signals on its dendrites, this can increase the amount of positive charge inside the body of the neuron. If that increase in positive charge increases beyond a certain threshold voltage, then the neuron will "fire", sending an action potential shooting down the neuron's axon to the next neuron. For the neuron, this thresholding effect is called the "all-or-none" law.[8]

In his book *The Emperor's New Mind*, the physicist Roger Penrose considers how this "all-or-none" law makes neurons behave as though they were elements of a digital computer:

> *An important feature of nerve transmission is that the signals are entirely "all-or-nothing" phenomena. The strength of the signal does not vary: it is either there or it is not. This gives the action of the nervous system a digital computer-like aspect.*

So both transistors and neurons fire when the combined inputs exceed a certain threshold voltage – and this is no coincidence. The threshold voltage is essential. In both cases, the threshold voltage introduces the required nonlinear behaviour – and it is that nonlinear behaviour which allows us to build complex systems (computers or brains) from transistors and neurons.

---

[8] http://tinyurl.com/allornonelaw

And now we see that possible reason why signals are sent down the axons of neurons using the rather surprising chemical method – instead of simply sending an electrical current down a "wire". The chemical method creates the crucial threshold voltage inside the neuron, and it is the existence of that threshold voltage which introduces the essential nonlinear behaviour. In fact, the peculiar chemical method inside the neuron might be considered as turning the neuron into a form of nonlinear "semiconductor", so to speak.

Paul Davies emphasises this nonlinear behaviour of neurons in his book *The Cosmic Blueprint*: "The electrical output signal of a given neuron will depend in a **nonlinear** way on the combined input it receives from its connected partners." (That book includes a good section on nonlinearity).

There is one simple message I want you to take from this discussion: in its design and nonlinear functionality, **a transistor can be considered to be an artificial neuron.**

Integrated circuits

The similarity between transistors and neurons continues when we consider how most transistors are used nowadays. The vast majority of transistors are micro-miniaturised onto a semiconductor substrate to form an *integrated circuit* ("silicon chip"). The latest fabrication techniques allow extraordinary densities of up to 25 million transistors on a square millimetre of silicon. This actually results in an individual transistor size which is rather smaller than a neuron, but it is clear that the principle of packing microscopic transistors onto an integrated circuit resembles the packing of microscopic neurons in a brain.

Here is an image of the Apple A11 microprocessor. If you have the latest iPhone then you are already in possession of one of these. This is an image of the entire external package – the actual "chip" and its 4.3 billion transistors is contained within it:

With 4.3 billion transistors in an integrated circuit, it is clear we are approaching the number of 100 billion neurons in the brain. So why do integrated circuits possess nowhere near the functionality and flexibility of the brain? The answer must surely lie in the extreme connectivity of the neurons in the brain: each neuron can be connected to as many as 10,000 other neurons. In order to achieve that level of connectivity, it would appear we will eventually have to move away from fabricating integrated circuits on a flat two-dimensional plane, and instead create three-dimensional cubes of circuitry – again, moving in a direction which resembles the structure of an actual brain.

Three-dimensional chips are now starting to appear. Samsung have recently released computer memory chips featuring 38 layers of transistors, and the rumour is that this is seen as the future of the industry. One website presents this development as a simple solution to the problem of how to achieve more of the crucial interconnects in integrated circuits, and compares the move to three-dimensions on a chip with the drive to build high-rise towers in cities:[9]

> *The obvious choice for where to put these interconnects was the same solution in any sprawling metropolis; if you can't grow* **out**, *grow* **up**.
>
> *Three dimensional chips will be released. It is only a matter of time. There is simply no other way to increase the density of interconnects, the number of devices on a chip, or speed than by moving into a third dimension of silicon.*

---

[9] *The Coming Age of 3D Integrated Circuits*, Brian Benchoff, **http://tinyurl.com/threedchip**

The development of integrated circuits which resemble the structure of the brain is called *neuromorphic computing*. It is currently an area of rapid growth. As an example, IBM released its True North integrated circuit in 2014 which contains over a million artificial "neurons" on a chip. Each neuron has 256 programmable "synapses" giving a total of just over 268 million synapses. The artificial neurons communicate using transient electrical spikes – just like real neurons. This chip has been described by IBM as "literally a synaptic supercomputer in your palm".[10]

It is clear that each successive innovation in the integrated circuit industry is moving us closer to creating true three-dimensional artificial-neuronal electronic brains.[11]

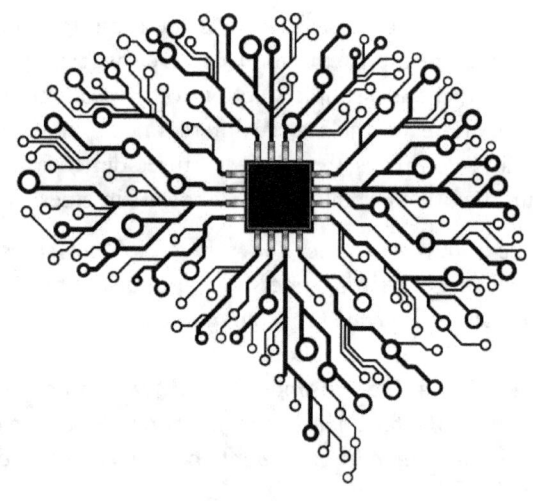

---

[10] "Introducing a Brain-inspired Computer", Dharmendra S. Modha, **http://tinyurl.com/truenorthbrain**

[11] Thanks to Alexander Bystritsky for the discussion and idea.

Digital information

The information inside most integrated circuits is purely digital. That means all information is represented in strings of binary "bits", each bit being either 0 or 1. The principle of Claude Shannon's "bits" as a means of representing information was described in Chapter One of this book.

Electrical input signals – coming into the metal legs of the integrated circuit – must also clearly represent either a 0 or 1 bit. If the specification sheet for an integrated circuit is examined, the corresponding voltage levels representing 0 or 1 are defined. The following diagram represents the actual acceptable input voltage levels for a family of common integrated circuits:

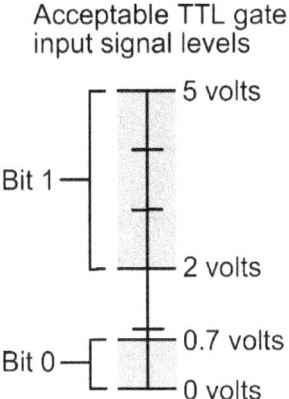

You can see from the previous diagram that the acceptable input voltage level for a bit 0 is from zero volts up to 0.7 volts. Why 0.7 volts? You will remember that 0.7 volts is the threshold voltage for a transistor. Any voltage lower than that level will not be sufficient to turn that transistor "on".

The threshold voltage on a transistor therefore performs a valuable function. Imagine an input signal coming into an integrated circuit with a value of zero volts (bit 0). If there is a temporary noise spike on that signal of anything less than 0.7 volts, that noise will not be sufficient to turn the first transistor "on" – the signal will still be interpreted as bit 0. So the threshold voltage acts to "clean up" the input signal and insulate the integrated circuit from external noise.

We now see how the threshold voltage in both neurons and transistors can act to convert a continuous, noisy, analogue signal into clean digital information. In *The Emperor's New Mind*, Roger Penrose describes how a similar thresholding mechanism is used by the cells in the retina of our eyes in order to reduce speckled noise in the images we see. Penrose explains that a single photon hitting the retina will not be sufficient to fire a signal:

> *In this case there is an additional mechanism present which suppresses weak signals, so that they do not confuse the perceived picture with too much visual "noise". A combined signal of about seven photons is needed in order that a dark-adapted human subject can actually become aware of their arrival. The type of cells that have been examined all require a threshold to be reached, and a very large number of quanta are needed in order that the cell will fire.*

Once converted into a digital format, the input data can be stored in our brains as digital information. Digital information does not decay over time: a digital photograph stored as a file on a computer does not fade like a physical photograph. Digital information can be transmitted without loss and degradation. In contrast, analogue data will inevitable degrade each time it is copied (try making successive photocopies of photocopies). Digital information seems insulated from physical damage. In similar fashion, our mental images are uncorrupted by any interference "snow" which used to afflict old analogue televisions. When we imagine old memories, there are no random white spots appearing in those images. Instead, our mental images seem noise-free like a modern digital television.

In Chapter One is was explained how a consciousness based on information would feel insulated from the physical world. Even though your brain is composed of a material with the consistency of rice pudding, you would not feel as though you were made of rice pudding – your consciousness would feel substrate-independent, independent from the physical world, and insulated from physical damage and the ravages of time. We now see how the threshold voltage – in both neurons and transistors – further acts to insulate our consciousness from the noise of the physical world. The thresholding mechanism prevents the accumulation of electrical noise, both in microprocessors and in our brains. It is as if our thoughts exist in a realm above the physical world, and insulated from physical damage.

4

COMPLEXITY

In the last chapter it was explained how a large number of nonlinear components can combine to produce systems which cannot be simplified to a simple component. This means that the behaviour of the multiple-component system could never be predicted from just studying the single component. In practice, it means that the behaviour of the multi-component system can be wildly different – completely unrecognisable, in fact – from the behaviour of the single component.

Multi-component nonlinear systems which behave in these surprising ways are said to exhibit *emergent* behaviour. Human intelligence and consciousness is typical of this emergent behaviour, emerging from the interactions of billions of neurons. You would never imagine that behaviour so extraordinary as intelligence and consciousness could arise if you just considered a single neuron in isolation.

The science which considers these nonlinear systems – and their emergent behaviour – is called *complexity*. We shall consider complexity in this chapter.

Chaos

Complexity – and its implications for science – was largely ignored by researchers until the second half of the twentieth century. The catalyst for complexity research was the invention of the digital computer. Suddenly, it became possible to simulate complex systems in the laboratory.

The first example of this surprising emergent behaviour in computer simulations came with the discovery of *chaos*. In the 1960s, the meteorologist Edward Lorenz working at MIT performed a very simplified simulation of weather systems on a computer and found that his simulation behaved in a surprisingly unpredictable manner. The long-term behaviour of the simulated weather displayed extreme sensitivity to initial conditions – any slight difference would eventually produce a wildly different outcome. Hence the well-known phrase about the chaotic behaviour called the *butterfly effect*: "A butterfly flapping its wings in China could cause a hurricane next month in New York". The implications for weather forecasting were clear, as stated on the UK Met Office website: "When looking at forecasts beyond five days into the future the chaotic nature of the atmosphere starts to come into play". It can never be possible to make an accurate weather forecast more than a week or two in advance – no matter how accurate are your measurements or how powerful your computer.

Chaos was considered in detail in my fourth book. In that book, I presented an example of chaotic behaviour emerging from the remarkably simple nonlinear mathematical expression $2x^2-1$ (it was explained in the previous chapter how the x^2 term results in nonlinear behaviour). Using that mathematical expression, it was shown how the following apparently random graph can be produced:

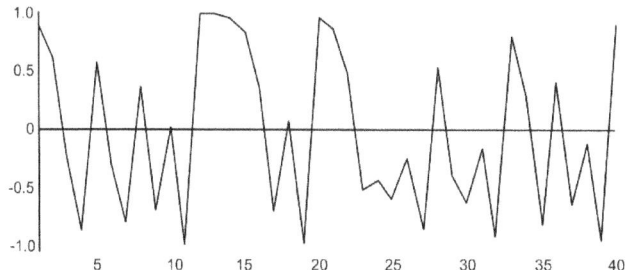

It is clear from the graph that dramatically random behaviour can be produced from very simple arrangements of a few nonlinear components. In fact, this is how pseudo-random numbers are produced in hand-held electronic calculators: a very small subsystem composed of transistors in the calculator's main integrated circuit can produce the random stream of numbers on demand.

With it being so easy to produce such dramatically random behaviour from a few nonlinear components, it is easy to imagine how a small module of chaotic circuitry in our unconscious brain could be responsible for our random creativity and flashes of inspiration apparently appearing "out of nowhere". There is surely no need to imagine that any more exotic processes – such as random quantum processes – would need to be involved to generate our creativity and inventiveness.

In the December 1986 issue of *Scientific American*, a major article on chaos described this possibility:

> *Innate creativity may have an underlying chaotic process that selectively amplifies small fluctuations and molds them into macroscopic coherent mental states that are experienced as thoughts. In some cases the thoughts may be decisions, or what are perceived to be the exercise of will. In this light, chaos provides a mechanism that allows for free will within a world governed by deterministic laws.*

Emergence

Chaos is one example of the surprising and remarkable behaviour which can emerge in complex systems which contain a huge number of elements. This sudden jump to a new mode of behaviour is called *emergence*, and the resulting behaviour would then be called "emergent" behaviour.

In this book, we are particularly interested in emergence as it is clear that consciousness is a form of emergent behaviour which results from the interaction of billions of neurons. John Barrow describes this surprising nonlinear jump in behaviour in his book *Impossibility*:

> *Complex structures seem to display thresholds of complexity which, when crossed, give rise to sudden jumps in the complexity. Take groups of people. One person can do many things; add another person and a further relationship becomes possible; but gradually add a few more people and the number of complex interrelationships grows enormously. Economic systems, traffic systems, computer networks: all exhibit sudden jumps in their properties as the number of links between their constituent parts grows. Consciousness is the most spectacular property to emerge in this way when a very high level of complexity is reached in a connected logical network, like the brain.*

A dramatic example of emergent behaviour is the termite mound, emerging from the behaviour of millions of termites. A termite mound is a vertical city, resembling a human skyscraper, and containing very similar heat regulation and air conditioning systems. The tallest structure constructed by humans is the Burj Khalifa Tower in Dubai, which stands 828 metres tall. With a man measuring – on average – 1.73

metres tall, that means the Burj Khalifa is 479 times taller than its human builder. In comparison, the tallest termite mound so far discovered is 12.8 metres high. With a single mound-building termite measuring just 0.6 cm in length, that means a termite mound is 2013 times larger than its builder.[12]

It is clear that termites have the edge when it comes to skyscraper construction. Here is a photograph of a man standing between two enormous termite mounds in the Northern Territory of Australia:

However, each individual termite has no knowledge of how to build the mound, and there is no single "foreman" in overall charge of the project. Each individual termite pursues a very simple operation, guided only by its local environment and its genetically-encoded rules. If you were to study the behaviour of just a single termite you would see no

[12] *Which species is the best builder: humans or termites?*
BBC website: **http://tinyurl.com/termiteskyscrapers**

indication that they were capable of such extraordinary feats. The mound only "emerges" when millions of termites interact.

This sudden and unexpected emergence of an impressive capability has seen the property of emergence being described as "The whole is greater than the sum of its parts". As an example, it would certainly appear that the property of consciousness is far more impressive than any property possessed by a single neuron.

Though not everyone is so impressed …

Reductionism vs. emergence

Many scientists would feel uneasy – or even reject outright – this notion of emergence, the notion that the behaviour of a system cannot be reduced to considering the behaviour of one of its component parts. That is because the philosophy seems to strike against the way we have done science for 400 years.

The modern scientific method is a relatively new phenomenon, the start of which can be traced back to Galileo in the 17th century. René Descartes described his own scientific method in 1638:

> To divide all the difficulties under examination into as many parts as possible, and as many as were required to solve them in the best way, and to conduct my thoughts in a given order, beginning with the simplest and most easily understood objects, and gradually ascending, as it were step by step, to the knowledge of the most complex.

The suggestion of Descartes is that the best way for science to progress is to break Nature into its smallest pieces, and analyse how those pieces behave. This has certainly proven to be a successful strategy for science, and has underpinned our technological advances.

This philosophy that a full understanding of behaviour can be achieved by considering the smallest elements of a system is called *reductionism*.

However, the scientist who believes in the importance of emergent behaviour would say the philosophy of reductionism is like saying: "I understand how a brain works because I understand how a neuron works", or "I understand how a computer works because I understand how a transistor works".

The difference of opinion between the reductionists and those who believe in the importance of emergence is often portrayed as an ideological conflict. So which is right: reductionism or emergence?

Well, as is usually the case when you have two sides with strong arguments, both sides are right – and both sides are wrong. It all depends on the behaviour of the components which make-up the system of interest. Specifically, it depends on whether the components which make-up the system are linear – or nonlinear.

The solution was actually presented in the previous chapter when we considered nonlinearity. In that chapter, it was explained how systems which were formed from linear elements could successfully be reduced to a single component while still maintaining the behaviour of the system. Remember the resistor network being reduced to a single resistor? That was a typical example of reductionism: the single resistor had exactly the same behaviour as the combined network of seven resistors. Reductionism is a convincing argument – with just one problem: **it is only true for linear components.** As soon as you shift to consider nonlinear components (such as neurons and transistors), the argument breaks down. As described in the previous chapter, a system which is composed of nonlinear elements cannot be successfully simplified: the behaviour of the composite system is not completely contained in the behaviour of a single component. This principle is described in the December 1986 issue of *Scientific American*:

> *The interaction of components on one scale can lead to complex global behavior on a larger scale that in general **cannot be deduced from knowledge of the individual components.***

By simplifying the system, you lose some aspect of the overall behaviour. In that case, your only option is to consider the large-scale emergent behaviour. On the plus side, that emergent behaviour due to nonlinearity can be very interesting indeed (consciousness, for example).

Particle physicists tend to be reductionists: they operate at the lowest-level, and they assume that systems will scale linearly. For many systems and materials, that will be the case: a sample of a few atoms of iron behaves in just the same way as a planet-sized block of iron, for example. If you know the behaviour of atoms, it is a simple process to scale that behaviour up to systems consisting of billions of atoms – as long as those atoms remain in a regular arrangement so that their behaviour scales linearly: a crystal, for example. It is believed that there is a 1,500 mile-long crystal of iron at the centre of the Earth, and geologists know how it behaves because they know how single atoms of iron behave: the behaviour scales linearly.

However, sometimes large numbers of atoms can combine to produce surprising behaviour which is highly nonlinear. The study of these materials is called *condensed matter physics*. Condensed matter physics is the most popular area of physics research (according to a recent poll, one third of all American physicists consider themselves to be condensed matter physicists). That means one third of American physicists are studying emergent behaviour. However, the field gets very little publicity compared to other areas of physics research. This general lack of awareness of the subject was made clear when the 2016 Nobel Prize in physics was awarded to condensed matter research. The announcement of the award sent science journalists scurrying to the internet to find out more about emergent behaviour.[13]

Another example of emergent behaviour when large numbers of particles interact is *superconductivity*. The phenomenon of superconductivity occurs when electrons act together and can freely move through a conductor as though there is no resistance. According to an article by the National Science Foundation:[14]

> *Take a metal such as lead or tin and cool it below a certain critical temperature: many of the individual electrons in that metal will suddenly start to march in step, so to speak – a collective motion that allows them to flow as if the metal offered no electrical resistance whatsoever.* ***Again, electrons in the aggregate exhibit behaviors that are nowhere to be found in one electron alone.***

---

[13] A profound article by particle physicist Jon Butterworth: "So from one point of view this physics looks more fundamental than the search for the smallest constituents of matter. It is the physics of deep principles, which these constituents seem to obey", **http://tinyurl.com/jonbutterworth**

[14] National Science Foundation, *Understanding Emergence*, **http://tinyurl.com/understandingemergence**

In the case of nonlinear components there is no way of predicting the onset of emergent behaviour unless you actually connect a huge number of the components together, or maybe run an equivalent huge computer simulation featuring possibly billions of simulated components. There are no short-cuts. As stated earlier, you cannot simplify large complex systems which result from the interaction of nonlinear components – the whole thing becomes a mathematical nightmare.

This is the reason why – as stated earlier – scientists only became able to tackle the problems of complexity when digital computers were invented. Maybe at this point you are starting to realise why complexity presents such a problem for conventional science.

I have seen this mathematical "nightmare" mentioned in the scientific literature. Melanie Mitchell is a professor of computer science who has worked at the Santa Fe Institute which is the world's leading centre for complexity research. In 2009, Mitchell wrote a book entitled *Complexity: A Guided Tour* which is an introduction and overview of complexity. Chapter Two of that book provides a clear explanation of nonlinearity. The mathematical "nightmare" that nonlinearity poses for reductionism is described by Mitchell:

> *Linearity is a reductionist's dream, and nonlinearity can sometimes be a reductionist's **nightmare**.*

It was mentioned earlier that emergence is often described as: "The whole is **greater** than the sum of its parts". In her book, Melanie Mitchell presented a similar succinct phrase to describe reductionism in which the behaviour of components scales in a linear and predictable manner: according to Melanie Mitchell, reductionism can be described as "The whole is **equal** to the sum of its parts".

A measure of emergence: Φ

In the previous chapter, it was explained how Christof Koch and Giulio Tononi have developed a theory of consciousness which is called integrated information theory, or IIT. This is regarded as being the current leading theory of consciousness. If you remember, IIT involves considering the connectivity of the network of information in the brain.

Giulio Tononi has described a numeric value which can be calculated from that connectivity of information. According to Tononi, this numeric value – which is called Φ (the Greek letter "phi") – then represents the consciousness of the network. According to Tononi, this calculation could be applied to anything: from an iPhone to the Milky Way, and the calculated value of Φ would reveal whether or not the object was conscious.

After reading more about Tonini's work, it seems rather clear to my mind that Φ is a measure of the emergent behaviour of a system: the more emergent the behaviour, the more likely the system is to be conscious. Indeed, in his book entitled *Phi*, Tononi describes Φ as the extent to which "the whole is more than the sum of its parts" – the definition of emergence.

The following discussion describes my attempt to visualize this measure of emergence: Φ.

IIT possesses a cool and imaginative feature which is called the "cruellest cut". Tononi considers the situation in which a network of information might be sliced into two smaller networks. What would be lost by doing that? Again, the question boils down to the distinction between reductionism and emergence: a system formed of linear components would lose nothing by being sliced in half, whereas a system formed of nonlinear components –

exhibiting strongly emergent behaviour – might find its performance severely reduced by being sliced in half.

The "cruellest cut" is specified as being the single slice of the network which has the most destructive effect, reducing the emergent behaviour by the greatest amount. To illustrate the principle of the cruellest cut, Tononi considers examples involving slicing a conscious brain in two, or cutting the head off a person's body – both of which, I would suggest, place this aspect of the theory firmly in the "untestable" category!

Let us consider the particular example of the brain. It has been described how a brain has approximately 100 billion neurons. We might slice a brain into two parts, each containing 50 billion neurons. In general, this most dramatic of actions – the cruellest cut – would significantly reduce the consciousness present in each of those half-brain parts. In fact, it might eliminate consciousness altogether.

So if 50 billion neurons represents "low consciousness" and 100 billion neurons represents "high consciousness", we might draw a graph showing how consciousness varies with the number of neurons:

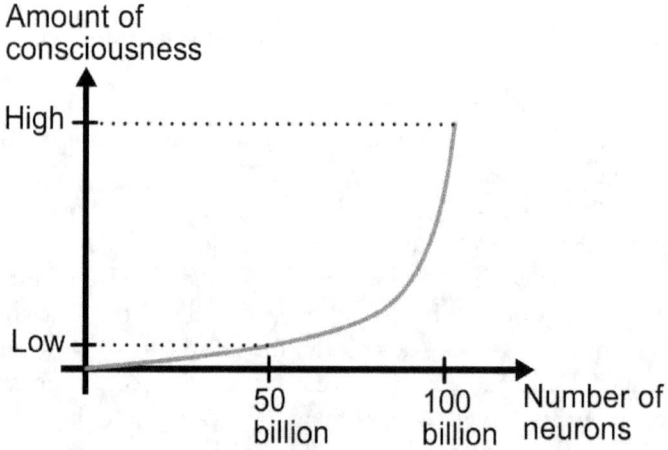

You will see from the previous graph that a part containing 50 billion neurons is shown as having "low" consciousness, whereas a part with 100 billion neurons is shown as having "high" consciousness. So the graph is highly nonlinear, with a pronounced upward kink at about the 90 billion neuron mark. This extreme nonlinearity is the signature of emergent behaviour.

So let us now apply that "cruellest cut", dividing the brain into two parts, each part containing 50 billion neurons:

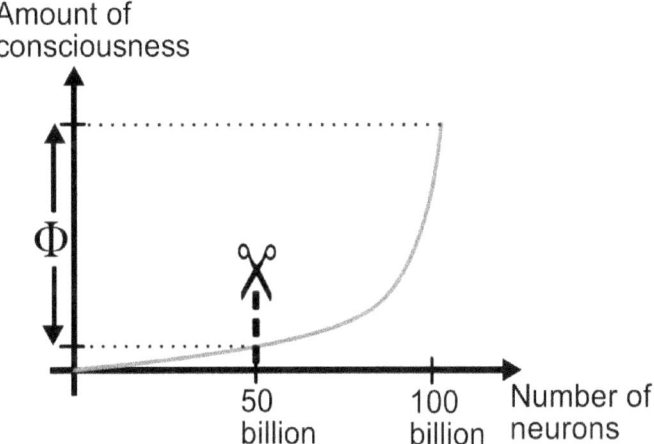

On the previous diagram, you can see a pair of scissors applying the cruellest cut, splitting the brain into two 50-billion-neuron parts. If the cut makes a big difference, then that reveals that the 100-billion-neuron network (the whole brain) vastly outperforms a 50-billion-neuron network (the half-brain part). That would be a clear sign of a highly-emergent network, a sign that "the whole is greater than the sum of its parts". As shown on the diagram, that would result in a large value of Φ.

In contrast, the following diagram shows a situation in which the behaviour of a system scales linearly, in other words, there is no sudden jump in behaviour, no emergent behaviour:

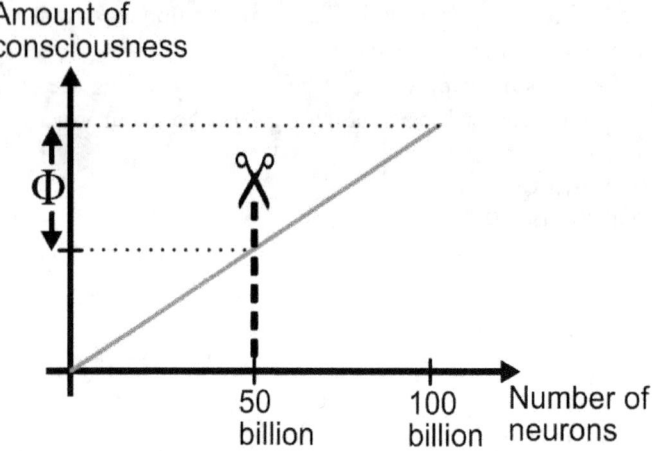

You will see on the previous diagram that the same cruellest cut has been applied at the 50-billion-neuron level. However, in this linear case, the destructive effect of the cut is reduced. Each 50-billion-neuron part has a greater amount of consciousness than the previous example. In this case, it could be said that the whole is **equal** to the sum of its parts – which, as has been stated, is the signature of reductionism. This system could be decomposed into smaller units with no loss of performance. As shown on the diagram, the measured value of Φ would be smaller in this case.

According to integrated information theory, we can then interpret the value of Φ as the amount of consciousness present in any network.

The low-hanging fruit

This seems like a good opportunity for me to say something which is possibly rather controversial …

On the basis of the discussion so far, we now see why reductionism has been so closely associated with the scientific method. Whenever scientific researchers have had the freedom to pick-and-choose their domain of interest, they have tended to concentrate their efforts on areas which are amenable to this method of decomposition into smaller components. In practice, that means the study of systems formed of linear components. Fortunately, as explained earlier, this has been sufficient to explain the behaviour of many materials when the behaviour of individual atoms is known. The discovery of semiconductors – and the resultant technological revolution initiated by their discovery – would be a good example of this: it has proven to be a route to success.

Unfortunately, this freedom of choice has meant that scientists have tended to avoid considering the difficult mathematically-intractable systems formed of nonlinear components – all of which explains why we have made the amazing discovery of the Higgs boson, while at the same time we are left almost completely in the dark about how the brain in own heads works.

Put more simply, areas have been picked in which most progress can be made, perhaps taking the easier option, and areas involving complexity have been avoided. Because of those choices, the suggestion might be made that scientists have picked the "low-hanging fruit".

The physicist Paul Davies has written a book about complexity called *The Cosmic Blueprint*. At the end of the following extended quote from his book, Davies refers to the general lack of interest in analysing nonlinear complex systems, and how this has resulted in them being "neglected" by scientists (I have placed the relevant sentence in bold):

The greater part of modern science and technology stems directly from the fortunate fact that so much of what is of interest and importance in modern society involves linear systems. Roughly speaking, a linear system is one in which the whole is simply the sum of its parts. Thus, however complex a linear system may be it can always be understood as merely the conjunction or superposition or peaceful coexistence of many simple elements. Such systems can therefore be decomposed or analysed or reduced to their independent component parts. It is not surprising that the major burden of scientific research so far has been towards the development of techniques for studying and controlling linear systems.

By contrast, nonlinear systems have been largely neglected. *In a nonlinear system the whole is much more than the sum of its parts, and it cannot be reduced or analysed in terms of simple subunits acting together. The resulting properties can often be unexpected, complicated and mathematically intractable.*

It is interesting to consider the distribution of complex systems according to their size, from the smallest objects in the universe to the largest. Galaxies are huge, but they can be treated as a single unit, rotating about a centre of mass. We can then describe the behaviour of a galaxy by using the relatively simple equations of general relativity. Similarly, at the smallest scales, elementary particles can be described by the relatively simple equations of quantum mechanics. The most complex objects – which therefore cannot be described by simple equations – are found in the mid-scale region.

In 2015, Neil Turok of the Perimeter Institute presented a lecture entitled *The Astonishing Simplicity of Everything* (available on YouTube). In that lecture, he considered how complexity is only found in the middle between the two extreme scales of size. And it just so happens that it is in that middle region of high complexity where we find life, and humans, and conscious brains:

> *We are in the middle in terms of scale. The astonishing thing about recent discoveries in physics is they tell us the universe is surprisingly simple and regular – on the tiniest scale, and on the hugest scale. It's only complicated in the middle.*

The following diagram shows the variation in the complexity of objects according to their size. Just as Neil Turok suggests, the objects with least complexity can be found at the extremes of the scale, whereas life (and the human brain – the most complex object we know of in the universe) is found precisely in the middle of the scale:

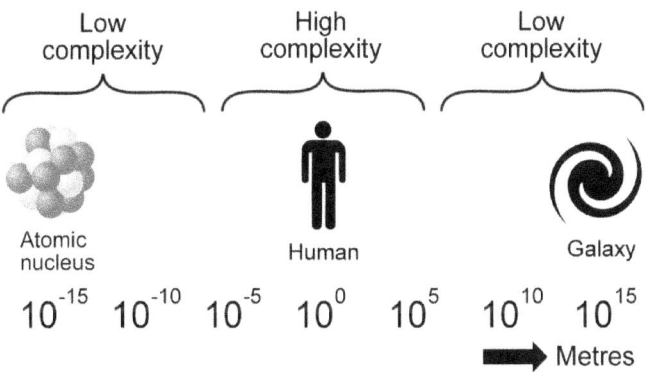

The field of physics has progressed by explaining puzzling phenomena by describing that behaviour in terms of simple formulas and laws. For example, the behaviour of the huge celestial bodies – the orbit of the Moon around the Earth, for example – can conveniently be described by nothing more than Newton's laws of motion and a few lines of mathematics. How convenient for physicists!

But imagine a situation in which the size of complex objects had extended out to the extremes of the scale – maybe with living creatures the size of stars or galaxies. In that case, such simplifying methods could not have been applied: we could not have accurately predicted the rotation of galaxies, for example. The behaviour of a galaxy would have, instead, depended on the complex behaviour of its "mind". If that had been the case, the field of physics would have struggled to advance, and the behaviour of the universe would have largely remained a mystery to us.

5

CAN A COMPUTER THINK?

It is the year 2021. The world has been ravaged by the nuclear World War Terminus. Rick Deckard is one of the survivors, struggling to get by on his income as a bounty hunter in a post-apocalyptic San Francisco.

Deckard is hired by the police department to "retire" (i.e., kill) androids which have returned from distant space colonies. But Deckard has a problem. The problem stems from the technology in the latest Nexus-6 androids. According to the spec sheet:

> *The Nexus-6 brain unit they're using now is capable of selecting within a field of two trillion constituents, or ten million neural pathways.*

As a result, the Nexus-6 androids ("andys") are so advanced that they are almost indistinguishable from humans. According to Deckard: "No intelligence test would trap such an andy". In conversation, there is no way of telling the difference between a Nexus-6 android and a human. Deckard has only one hope: the Voigt-Kampff Empathy Test. Apparently, androids feel no empathy with others of their kind: "Empathy, evidently, existed only within the human community". Deckard has to rely on this emotional deficit to detect the rogue Nexus-6 machines.

By now you might have realised that I am describing the plot of Philip K. Dick's classic science fiction novel *Do Androids Dream of Electric Sheep?* The book later formed the inspiration for the equally-classic Ridley Scott movie *Blade Runner.*

Let us rejoin the story of Rick Deckard as he rides in his flying car to the headquarters of the Tyrell Corporation where the androids are manufactured …

At the Tyrell Corporation, Deckard meets Tyrell's assistant, Rachael. After performing the Voigt-Kampff test on Rachael, he discovers she is an android. However, Rachael is clearly intelligent, conscious, and alluring. What is more, Rachael has been given false memories so she does not realise she is an android.

After a while, Deckard realises he has started feeling "empathy" for Rachael – or maybe something more: "I'm capable of feeling empathy for at least specific, certain androids. Not for all of them but – one or two". Deckard is now faced with a dilemma, and has to wrestle with his conscience: is it ethical to "retire" Rachael, another seemingly-conscious being, even though she is an android?

Deckard now starts to question if he is in the right job: "Suddenly, for the first time in his life, he had begun to wonder."

What should Deckard do? How should he behave toward Rachael?

At the heart of the problem appears to lie the question of whether or not an android can be conscious. As described in Chapter One, we use the fact that we are conscious to separate ourselves from the everyday objects – such as clocks and cars – which we do not believe are conscious. When faced with a potentially-conscious android, we would lose that certainty that we are superior. It seems it is the presence of our consciousness which defines us as human.

The need to resolve these questions about the consciousness of androids will become urgent as androids start to perform working roles. For example, it appears androids – combined with the latest speech recognition artificial intelligence – will soon be acting as multi-lingual airport and hotel receptionists. Here is a photograph from a Tokyo convention of the latest DER2 android produced by the Kokoro Corporation. The android responds to commands in Japanese, Korean, Chinese, and English:

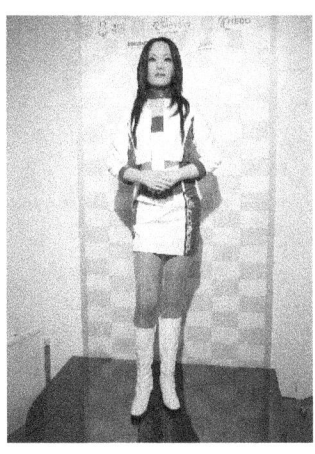

A video of the DER2 in action is available on YouTube:

http://tinyurl.com/replicantvideo

Developments in artificial intelligence in androids might uncover a legal minefield. Would a physical assault against one of these androids be regarded as a serious crime, as serious as an assault against a human worker? Should it be considered to be a crime to turn off the power to one of these androids, thereby "killing" it?

If you read the video description on YouTube for the previous video, you will find that at one point in the DER2 android demonstration the android had to warn men not to touch her as that would be considered sexual harassment. The men who were watching the demonstration treated the request seriously and apparently nodded their heads, agreeing to comply. The men had already decided that the android deserved to be treated with equal respect.

And, when Rick Deckard finally decided how to treat Rachael, he came to the same conclusion.

It seems that the answer to these difficult questions will be decided by whether or not we determine the android to be conscious. So in this chapter we will be considering the crucial question as to whether a computer (and, thereby, an android) could ever be conscious. We will be moving away from the hard reality of neurons and transistors, and entering the abstract world of computing and mathematics.

In Chapter One, the principle of substrate-independence was described. That was the idea that if the neurons in your brain were replaced with transistors – or any other form of similar hardware – you would still be conscious. If we assume the principle of substrate-information is true, then that removes one major fundamental objection to a computer becoming conscious.

However, another objection has been raised, based on the method by which modern computers "think" to solve problems. At the heart of a modern computer lies a *microprocessor* (an integrated circuit) which performs one logical or mathematical operation at a time. Problems are then solved by following a step-by-step computer program (or *algorithm*). Each step of the algorithm is clearly-defined, allowing no ambiguity. Every step of the process is as strictly defined as the most formalised piece of mathematics.

However, it has been suggested that this highly-formal approach to problem solving is not a good match for the human brain as the approach appears to allow no room for ingenuity, random creativity, or intuition. Do humans really follow a rigidly strict sequence of instructions every time they want to solve a problem? Was that really how Einstein's thought processes worked? I very much doubt it.

In another criticism of the algorithmic approach, we will see in this chapter how Alan Turing discovered a remarkable result which revealed that there were some things computers could never compute. Just like the empathy-deficit of the Nexus-6 android, does that mean there will inevitably be fundamental deficiencies in the processing of a computer mind compared to the human mind? Specifically, does all this mean that a computer can never be conscious?

In this chapter we will examine if there are any fundamental limitations on a computer's thought processes. We will start by considering a very dangerous entity for any inflexible and strictly-logical system: the danger of the paradox.

The liar's paradox

In the early decades of the 20th century, the foundations of mathematics were threatened because of the recent discovery of paradox lurking within mathematics. The paradox was based on the *liar's paradox*. As an example, if someone says: "I am lying" or, equivalently "Everything I say is false" then that would represent a paradox. The statement can be either true or false. If the statement is true, then – according to the statement – the statement is false (because the person is lying). Alternatively, if the statement is false, then – according to the corrected statement – the person is telling the truth, so the statement as told by the person must be true.

Either way, we appear to have a statement which is self-contradictory: if it is true then it is false, and if it is false then it is true. That constitutes a paradox.

The particular form of the liar's paradox which threatened the foundations of mathematics is called *Russell's paradox*, and it was discovered by the English philosopher and mathematician Bertrand Russell in 1901. Russell's paradox concerns *sets*, which is the mathematical term for a collection of objects. The paradox can perhaps be easiest understood if we consider an example. So let us consider the collection of men who are shaved by a barber. That collection of men would, mathematically, be known as a set.

To understand how the paradox can arrive, consider the situation in which the barber shaves a man if and only if the man does not shave himself. Let us consider the set of men who are shaved by the barber.

So let us now ask the question: is the barber himself in the set? In other words, does the barber shave himself or not?

If the barber **is** in the set then, by the set definition, that means he **does not** shave himself. In which case then, according to his rule, he **does** shave himself and should therefore **not** be in the set. But if the barber is **not** in the set then, by the set definition, that means he **does** shave himself. In which case then, according to his rule, he **does not** shave himself and so **should be** in the set.

So, as with the liar's paradox, we have a situation which if it is true then it is false, and if it is false then it is true.

Paradoxes are so destructive in mathematics because mathematics attempts to clearly and unambiguously decide the truth or falsity of a statement expressed in clear mathematical language. As a simple example:

"Two plus two equals four"

is an example of a **true** mathematical statement, whereas:

"Two plus two equals five"

is an example of a **false** mathematical statement.

In that sense, mathematics can be seen as the "route to truth and clarity" about the world. And that truth is unchanging for all time: when a mathematical theorem is proved to be true, it remains true forever. However, if there are paradoxical statements which can be expressed mathematically which are **both** true **and** false at the same time, then that clarity becomes lost. We are left with indecision and uncertainty, and mathematics as some "route to truth" appears doomed.

An example of the destructive capability of a paradox was shown in an episode from the original series of *Star Trek* called "I, Mudd", broadcast in 1967. In the episode, the Enterprise is hijacked by an android named Norman. Norman reveals himself to be inflexibly logical in his thought processes, and Captain Kirk realises this represents a weakness which they can exploit. As a result, the android is defeated by the liar's paradox by simply telling it: "I am lying". The android quickly realises the paradox, and it proves to be more than the ultra-logical android brain can handle.

Here is the android's response to the paradox:

*You say you are lying but, if everything you say is a lie then you are telling the truth. But you cannot tell the truth because everything you say is a lie. But you lie … you tell the truth … illogical … illogical … please explain … you are human … **only humans can explain the behaviour …***

At which point smoke pours out of the android's head and it is destroyed.

The scene in which the android is destroyed is available on YouTube: **http://tinyurl.com/liarsparadox**

This represents an excellent demonstration of the destructive capabilities of a paradox, capable of tearing-down a strictly logical system. This also reveals how paradoxes were perceived as being such a threat to the logical foundations of mathematics at the start of the 20th century.

But the parable of Norman the android also appears to reveal a possible limitation on the "thought processes" of a computer. The final line spoken by Norman is particularly revealing: "Only humans can explain the behaviour". Norman appears to be admitting that there are fundamental differences between the workings of a human mind and the workings of a computer mind. It would appear that the human mind has a greater capacity to understand certain difficult concepts (for example, understanding the principle of a paradox). It would appear that a rigidly-logical computer would be unable to "get its head round" the concept of a paradox. However, humans (and Vulcans) would have no such limitation. Even the highly-logical Spock is able to consider the implications of the liar's paradox without smoke pouring out of his head.

So, can a computer think? Maybe, but it appears we have discovered that there might be limitations on what they can think.

The halting problem

The year is 1936.

Alan Turing – the brilliant English mathematician – is 22 years old and has just been awarded a research fellowship at King's College in Cambridge University. The fellowship was worth £300 a year, which was not much money even in those days, but it was enough to provide Turing the freedom to pursue his ideas in an idyllic environment.

Here is a photograph of King's College in Cambridge:

Turing was lying in the grass in Grantchester Meadows when he had a brilliant insight.[15] It was an insight which would reveal a truly extraordinary limitation on the ability of all computers.

---

[15] Grantchester Meadows are located about a mile outside Cambridge. "Grantchester Meadows" is also the name of a song by the Cambridge-based band Pink Floyd, whose founder members Roger Waters and Syd Barrett happened to be taught by my father in Cambridge High School.

Turing's idea considered the apparently simple question of determining whether or not a computer program eventually halts (stops running, and then usually producing some useful output). The alternative would be that the program might run forever, trapped in a never-ending loop, never halting to produce useful output.

The subject of Turing's idea – whether or not a computer program halts or loops forever – might appear quite abstract and uninteresting, but the consequences were to be huge.

It might appear fairly obvious that one way of determining whether or not a program halts is just to leave it run and see whether or not it eventually halts. However, we shall now see that that strategy will not work in all circumstances.

If you are on Facebook then you are probably tired of those silly mathematics problems that people share which turn out to be incredibly simple. The following image resembles one of those problems – in appearance at least. Let us call it the "Facebook fruit problem" as it features apples, bananas, and pineapples:

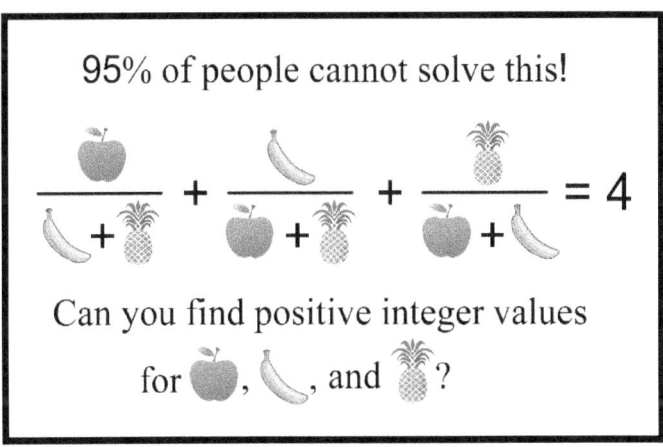

In order to solve the puzzle, you will see that you need to find the positive integer values represented by the apple, the banana, and the pineapple.[16] An example of the sort of whole-number solution we are seeking would be for the apple to have the value 5, the banana to have the value 3, and the pineapple to have the value 2 (these are not the correct answers). You might like to try to solve the problem yourself, but I warn you not to spend too much time on it.

You might write a simple computer program to solve this puzzle via trial-and-error. The program would start by setting the values of the apple, banana, and pineapple all equal to 1. The program would then test if those values solved the equation. If the equation was not solved, then the program would increase the value of one of the fruits by 1 and try again. The program would continue increasing the values until a solution was found. This would represent an exhaustive brute-force search of all possible solutions.

I can tell you that after a few years – or even a few decades – your program would not have found the solution to the puzzle. Therefore, after that length of time, and testing so many possible solutions, you might feel justified in stopping your program and announcing: "The puzzle has no solutions".

However, you would not be justified in halting the program and making that announcement. This is because, rather staggeringly, the correct value for the apple is:

154,476,802,108,746,166,441,951,315,019,919,837,485,
664,325,669,565,431,700,026,634,898,253,202,035,277,999

---

[16] These type of equations which have whole number solutions are called *Diophantine equations*. They are often used in puzzles, as in the Facebook fruit problem.

which is an astronomical number composed of 81 digits which is approximately equal to the number of atoms in the universe. It is approximately equal to ten thousand quadrillion quadrillion quadrillion quadrillion quadrillion.

Fortunately, the correct value for the banana is rather smaller. It is only:

36,875,131,794,129,999,827,197,811,565,225,474,825,492, 979,968,971,970,996,283,137,471,637,224,634,055,579

a number which merely contains 80 digits.

Finally, the correct value for the pineapple is relatively tiny, being composed of only 79 digits:

4,373,612,677,928,697,257,861,252,602,371,390,152,816, 537,558,161,613,618,621,437,993,378,423,467,772,036

If you did manage to solve the problem then, very well done indeed! Apparently, according to the Facebook meme, you can now consider yourself to be in the top 5% of the population (I'm joking – it means you probably work for the NSA).[17]

It would take longer than the lifetime of the universe to discover those whole numbers using a brute force method. It is therefore clear that the strategy of determining if a program eventually halts just by letting it run for a very long time will not be an effective strategy in all cases. You could never be justified in calling a halt to the program. You could never be certain that if you had just let the program run for a bit longer then it would have found a solution.

So we need to find a better method.

_____

[17] A technical discussion of this amazing Facebook fruit problem by Alon Amit is available here: **http://tinyurl.com/toughpuzzle**

In his book *Programming the Universe*, Seth Lloyd makes the point that sometimes it is very easy to tell if a computer program will halt, or loop forever, just by looking at the program. As an example, the simple one-line computer program:

PRINT "HELLO"

will just print "HELLO" and then halt (stop). So it is very clear to see that this simple one-line program will halt – without actually having to run the program.

As an example of the second type of behaviour, it is clear that the following two-line BASIC program will loop forever and never halt:

10 PRINT "HELLO"
20 GOTO 10

In the program, you will see at line number 10, the computer will print "HELLO". The computer will then advance to line 20 which says "GOTO 10", which means the computer has to loop back to line 10 and print "HELLO" again. This looping will continue forever (rapidly filling the screen with "HELLO"). It can be seen that this looping behaviour means that the program will never halt.

So it would appear it is possible to determine whether or not a program will eventually halt just by looking at the program and analysing that program – without actually having to run the program. We might then wonder whether we might be able to automate that process: could we write a sophisticated computer program which takes any other program as its input and then determines whether or not that input program will halt? Such a sophisticated program would be able to solve the Facebook fruit problem instantly – without having to run a brute-force algorithm for billions of years. The question of whether or not it is possible to write that general program is called the *halting problem*. The solution to the halting problem is not obvious, and it was considered to be one of the most important problems in early 20th century mathematics.

The halting problem caught the attention of Alan Turing, and it was while he was lying in the grass in Grantchester Meadows that the solution to the problem came to him. To cut a long story short, Turing was able to show that it would never be possible to write a single general computer program which would be capable of determining whether or not another program halts.

Turing's method was ingenious, and the form of the method resembles the "liar's paradox" which was described earlier in this chapter.

Firstly, let us imagine that it is possible to write a general-purpose algorithm to determine whether or not an input program halts. Let us denote that piece of code – which tests whether or not a program halts – by the following diamond in a flowchart diagram (the diamond represents a decision box):

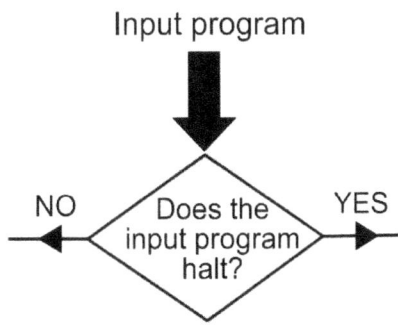

Turing proposed that that piece of code – which performs the halting test – could then be included as part of the following program:

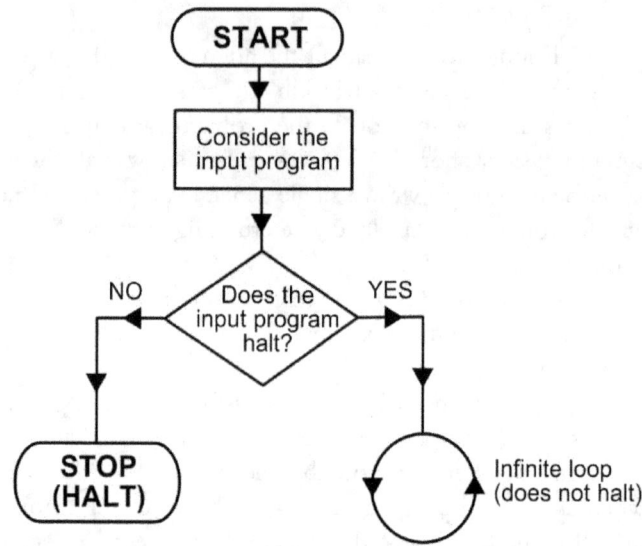

You will see from the flowchart that if the input program **does not** halt, then the flowchart halts. But if the input program **does** halt, then the program loops forever, i.e., it never halts.

So we can see from the flowchart that the behaviour of the flowchart algorithm is precisely the opposite of the behaviour of the input program. If the input program **does not** halt then the flowchart algorithm **does** halt. But if the input program **does** halt then the flowchart algorithm **does not** halt. As I said, the behaviour of the flowchart algorithm is precisely the opposite of the input program.

Alan Turing then did something ingenious. He took the flowchart algorithm (which can be coded as a computer program) **and presented it as input to itself!** Turing then asked the question: "Does the resulting program – given itself as input – halt or loop forever?"

There are two possibilities. If the input program **does** halt then – according to the algorithm – the flowchart algorithm **does not** halt and so the input program does not halt. Conversely, if the input program **does not** halt then – according to the flowchart – the flowchart algorithm **does** halt and so the input program does halt.

So if the input program does halt then it does not halt, and if the input program does not halt then it does halt. As in the liar's paradox, the self-referential nature of Turing's method has created a paradox.

A paradox is a situation which cannot exist: a computer program has to either halt or not halt – it can't do both. The only conclusion we can make is that the general halting tester program – the piece of code in the diamond – cannot exist. This method which Turing used in his proof is called "proof by contradiction".

Turing had therefore solved the question of the halting problem: it could never be possible to write a general algorithm which could determine if any other program either halted or looped forever.

Turing published his solution in 1937 in *Proceedings of the London Mathematical Society*. This was a magnificent achievement by Turing, and it established his reputation as a top-rank mathematician.

The implications for the android brain, however, are not so appealing. The result suggests that if we have a computer brain then there will always be a problem (the halting problem being an example) for which it is known that there is a solution (a program will definitely either halt or not halt) but the computer brain cannot calculate that solution. The solution is then said to be *uncomputable*.

The new mind of Roger Penrose

Turing's solution to the Halting Problem seems to suggest that there is a fundamental limitation of the computer mind. We will now see how one of the most eminent English physicists suggested that this limitation meant that computers do not function like the human brain, and therefore can never be conscious.

Roger Penrose's greatest achievement in physics came in the 1960s when he proved that a collapsing star had to form a singularity: a black hole. Working with Stephen Hawking, he then showed that the same principle had to apply to the universe as a whole: if you reverse time, the universe had to have emerged from a singularity at the time of the Big Bang.

In 1989, Roger Penrose wrote a physics book aimed at a general audience called *The Emperor's New Mind*. The market for popular science publishing in physics had been established a year earlier by Stephen Hawking's bestseller *A Brief History of Time*. As a result, Penrose's book was tremendously popular. I read the book when it came out as did many of my friends, and I know it motivated many readers to get interested in physics.

In his book, Penrose presented his case that computers operated fundamentally differently from the human mind and, as a result, computers can never be conscious. Penrose used Turing's solution to the Halting Problem which showed – as has just been described – that there are some problems which have solutions which can never be calculated by a computer. This appears to represent an intellectual deficiency of the computer – but it is not believed that humans possess a similar deficiency. For that reason, Penrose's theory suggested that these computer minds could never be conscious.

As stated earlier, computers solve problems by following a scripted algorithm in a very rigid, predicable, step-by-step manner. Penrose calls this type of behaviour *algorithmic*. An algorithmic mind has to stick to the script of its program, unable to go "off-script", being bound by rigid logic. An example of an algorithmic mind would be Norman the *Star Trek* android considered earlier. Norman's algorithmic mind is so rigid and inflexible that he is unable to consider the liar's paradox without smoke pouring out of his head.

In contrast, the minds of Captain Kirk and his crew are more flexible, not bound to following a step-by-step algorithmic program. As suggested at the start of this chapter, surely Einstein's mind did not follow a strict sequence of rules. Human minds are capable of going "off-script" by introducing original thinking. Penrose calls these types of minds *non-algorithmic*.

Penrose's theory is not fully stated until the final chapter of his extensive and wide-ranging book. As Penrose says in that final chapter: "It has, indeed, been an underlying theme of the earlier chapters that there seems to be something **non-algorithmic** about our conscious thinking." Penrose then introduces his theory which suggests a clear division between conscious and non-conscious minds on the following basis: **a non-algorithmic mind can be conscious, whereas an algorithmic mind can never be conscious.**

Therefore, according to Penrose's theory, we should not consider Norman the android to be conscious because his mind is algorithmic, forced to follow a strict program. In similar fashion, according to Penrose, no computer which follows a strict program could ever be conscious.

But how could we ever test Penrose's theory? It would appear we would need to compare the mind of a computer with the mind of a human, and somehow determine whether the algorithmic "mind" of the computer was conscious. But how could we ever get "inside the mind" of a computer to test Penrose's theory? It would seem to be an impossible task.

However, we have the perfect testing ground on which we can test Penrose's theory. And that testing ground is the human brain.

The brain appears perfect to test Penrose's theory because it is a device which is divided into two sections: one section operates algorithmically, while the other section operates non-algorithmically. What is more, we are able to "get into the mind" of this device, to determine if either of the two different sections are conscious.

As described in Chapter Two, the lower levels of the human brain are responsible for repetitive behaviours, such as walking and breathing, and also being responsible for monitoring body temperature and carbon dioxide levels in the blood. These are all functions which could be performed by a computer following a strict program. For that reason, we can consider the lower levels of the brain as representing **algorithmic** behaviour. As an example, the mind of Norman the android would operate like the algorithmic lower levels of our brain.

In contrast, the higher level of the brain does not follow a strict "program", and is therefore capable of going "off-script" to introduce original thinking, creativity, and problem solving for which – as Penrose suggests in the final chapter of his book – "no clear algorithmic process exists".

So let us redraw the simple map of the brain from Chapter Two to show the algorithmic and non-algorithmic regions:

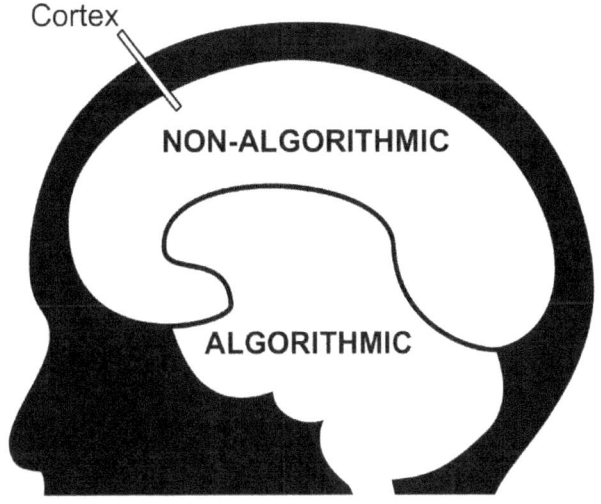

Let us remind ourselves what Penrose's theory suggests. According to Penrose, an algorithmic mind (like Norman the android) **cannot** be conscious, but a non-algorithmic mind **can** be conscious. And when we consider the two sections of the human mind shown in the previous diagram – **this is exactly what we find!** As described in Chapter Two, conscious behaviour is believed to reside in the higher level of the brain (which is shown to be the non-algorithmic region on the diagram), whereas the lower levels of the brain – the algorithmic regions – are believed to be unconscious.

To my mind, this represents strong evidence in support of the theory of Roger Penrose.

In the final chapter of his book, Penrose presents an description of how the conscious (non-algorithmic) and unconscious (algorithmic) mind work together to learn a new skill, for example, learning to drive a car. During the learning process, the conscious mind would be very active, solving problems ("How do I overtake the car in front?"). As

Penrose explains, the conscious mind would have to be used to solve the problem rather than the unconscious mind because "no clear algorithmic process exists" at that stage. However, once the problem is solved and you have overtaken a few cars, you have learned how to perform the task. It might be said that the learning process involves "writing your own computer program" to solve the task: you are generating a new algorithm. But once your conscious mind has generated that new algorithm, it can then be handed-down to the lower-level unconscious brain. It can then be said that you have learned a new skill. Then, the next time you need to overtake a car the unconscious brain can follow that program algorithmically and perform the task without having to involve the higher-level conscious mind. As a result we can drive our cars virtually automatically, without having to get consciously involved.

In his book *Consciousness*, Christof Koch presents a similar description of the difference between the conscious and unconscious mind. Koch explains how consciousness is needed for training, but then the unconscious mind can take over:

> *Much of the ebb and flow of daily life does indeed take place beyond the pale of consciousness. This is patently true for most of the sensory-motor actions that compose our daily routine: tying shoelaces, typing on a computer keyboard, driving a car, returning a tennis serve, running on a rocky trail, dancing a waltz. These actions run on automatic pilot, with little or no conscious introspection. Whereas consciousness is needed to learn these skills, the point of training is that you don't need to think about them anymore; you trust the wisdom of your body and let it take over.*

So it appears we might have found a valuable principle on our quest to uncover the secrets of consciousness: a clear definition of when biological material should be considered to be conscious, and when it cannot be capable of consciousness. If that material is algorithmic – if it follows a strict program – it cannot be conscious. But if that biological material is capable of non-algorithmic behaviour, then it has the potential to be conscious.

Quantum theories of consciousness

The latest incarnation of Roger Penrose's theory of consciousness is controversial because of Penrose's decision to incorporate some highly-unorthodox ideas based on quantum mechanics. This approach has received a great deal of criticism. So, in this section, let us consider some of the various quantum theories of consciousness.

Quantum theories of consciousness have proven popular almost from the first day that quantum mechanics was discovered. I would say there are two ways in which quantum mechanics might have an influence on theories of consciousness.

Firstly, quantum mechanics seems to present a mechanism by which the notion of "free will" might be given some foundation in science. When Newton proposed his three laws of motion in 1687, it appeared that the evolution of the universe was governed by strict laws – with no room for deviation. All successive states of the universe would be set in stone, determined by the original state of the universe and the application of Newton's laws. This principle has been called the "clockwork universe".

However, as the atoms of the brain follow Newton's laws, this would appear to indicate that the evolution of our thought processes – for our entire lives – are also determined by the position of the atoms in our brains at birth, and the application of Newton's laws. Where, then, is there any room for free will – the notion that we could behave differently if we chose to? Is free will just an illusion? Are all our decisions for our entire lives set in stone at birth?

The idea that all our decisions are actually decided well in advance was reinforced by a classic experiment. In the 1980s, a consciousness researcher named Benjamin Libet performed a series of experiments at the University of California, San Francisco. It was known that before a human subject performs a simple physical activity – such as pushing a button – there was a characteristic pattern of electrical activity in the brain known as the *readiness potential*. Libet set out to time this readiness potential, so he connected a volunteer to an EEG and the volunteer was asked to push a button several times at will. The volunteer had complete choice as to when he or she should push the button.

What Libet discovered was amazing. The readiness potential could be detected in the unconscious region of the brain about half-a-second before the person took the conscious decision to press the button. It appeared the decision to push the button was actually being made by the unconscious brain, with the apparently conscious decision being a mere illusion of free will occurring several milliseconds later.

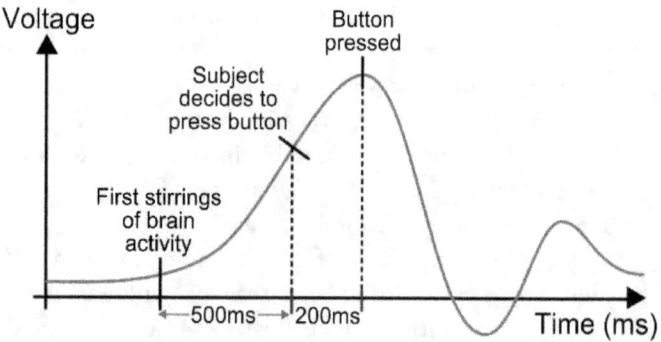

Libet's experiment appeared to reinforce the idea that there can be no free will in a clockwork universe, with everything being determined some time previously. However, with the discovery of quantum mechanics in the early 20th century, the door was opened for free will to make a comeback.

Quantum mechanics introduced fundamental uncertainty, and eliminated the concept of the clockwork universe. According to quantum mechanics, when we make a measurement at the most fundamental level, i.e., the level of individual particles, **we will get a fundamentally random result.** As described in my first book, the situation resembles a ball spinning round a roulette wheel. It is fundamentally impossible to predict in which slot the ball will land: you are prohibited from dismantling the system in order to predict the motion of the ball in terms of more fundamental quantities (because there are no more fundamental quantities – we are already at the fundamental level).

So, if the motion of the atoms in our brains are actually controlled by the random processes of quantum mechanics, this eliminates the concept of the clockwork universe and reintroduces the possibility of some degree of free will. This principle was described by Christof Koch in an interview which is available on YouTube:

http://tinyurl.com/freewillvideo

In the video, you will note that the interviewer says: "If our feeling of authorship is just a froth that arrives after our brain has made the actual decision then how can we possibly have free will?" in a reference to the previously-described Libet experiment.

However, if our thoughts are decided by the fall of random quantum dice, does that really mean we have free will in the conventional sense of the term?

The second proposed role for quantum mechanics in consciousness comes during the so-called "collapse of the wavefunction". In quantum mechanics, it is the case that particles can behave as though they are in a peculiar multi-valued state called a *superposition state*. For example, in the famous double-slit experiment a particle can appear to travel through two different slits at the same time – as though the particle is in two places at once. At that point, the particle is described mathematically by a *wavefunction*. However, once the particle has passed through the two slots and hits a screen, the particle is found to exist in only one place. In other words, when the particle is observed, it is said that the wavefunction has "collapsed" into a single state, and only one position of the particle is left.

But what physical process is responsible for this "collapse of the wavefunction"? In 1932 the mathematician John von Neumann – who was the foremost mathematician in the world at that time – proposed a controversial solution. Von Neumann wrote an influential book titled *The Mathematical Foundations of Quantum Mechanics,* and in that book he proposed that the collapse was caused by the consciousness of the experimenter. Basically, the experimenter formed the final link of a causal chain: by the time the experimenter considered the experiment, it was essential that the particle was only in one position. So the experimenter's final act of conscious observation was responsible for collapsing the wavefunction.

Von Neumann's "consciousness causes collapse" interpretation is not popular with physicists today, though in a recent poll, 6% of physicists still agree with it. However, our knowledge of the underlying physical principles of quantum mechanics has advanced since Von Neumann's time. It is now believed that the influence of the general environment (for example, dust particles hitting an electron, or the random arrangement of atoms in a screen hit by a particle) has the effect of "collapsing the wavefunction".

This is because, in general, environmental effects are random and noisy, so when a particle interacts with billions of random atoms in the environment, there is an "averaging" effect which progressively reduces the strange superposition effects and only leaves us with a well-defined state which is not in a superposition. This is why we do not find Schrödinger's famous cat both alive and dead at the same time!

This process by which the environment acts to "collapse the wavefunction" is called *environmental decoherence* or just *decoherence*. This process is now generally-accepted and uncontroversial among most physicists. The effect has been tested in many experiments, and, as we shall soon see, an understanding of the decoherence process is allowing new forms of powerful computing to be developed.

So there is no need for a conscious observer to "collapse the wavefunction". Any random environment can act as an "observer". For example, an electron hitting a screen will be "observed" by the screen, thereby revealing the position of the electron. The fact that a conscious observer is not required to collapse the wavefunction is described by Max Tegmark in his book *Our Mathematical Universe*:

> *Now I was convinced that consciousness had nothing to do with it, since even a single particle could do the trick: a single photon bouncing off a object had the same effect as if a person observed it.*

Because decoherence is such a powerful effect, usually occurring in a tiny fraction of a second, it is extremely difficult to keep a particle in a superposition state. However, if we can manage to keep a particle in a superposition – by completely insulating it from any interaction with the environment – then there might be big advantages. As an example, imagine a computer made out of particles in a superposition state. Just as particles in a superposition state can be in many places at once, so a computer made out of those particles would be able to perform many calculations at once! This is the general principle behind *quantum computing*.[18]

When he had finished writing his book *The Emperor's New Mind*, Roger Penrose was convinced that some form of non-algorithmic computing was necessary for consciousness, but he was unsure at the time where to find this new form of computing. But when he became aware of the extraordinary potential of quantum computing, this represented the possible breakthrough he was seeking. His feelings at the time are described in the following video (with Penrose still in great form at the age of 82):

http://tinyurl.com/penrosetheory

As Penrose states in the video:

> *To me, there is something outside the computational laws of physics. When I wrote my book The Emperor's New Mind I was trying to develop this idea, and I was trying to say there was something else out there. What could it be? Where is the biggest gap in our understanding of physics? There is a big gap within present day understanding of quantum mechanics.*

Penrose then wonders if "inside our heads we are **exploiting** that gap. That would take us to a level somewhat beyond present technology in our experiments". Penrose is inferring that it might be possible to have something resembling a quantum computer – with its associated

---

[18] In practice, only a quantum computer's **data** would be held in a superposition state.

extraordinary computing power – hidden somewhere in our brains. Up to this point, this all sounds reasonable enough. But it is at this point that the problems really start. Because, as Penrose states, in order to "exploit" quantum computing in our brain would require techniques "beyond present technology". Present technology is revealing just how difficult it is to keep particles in superposition states for extended periods of time – and avoiding environmental decoherence. Rudimentary quantum computers are now appearing, but the particles have to be isolated from the environment and kept at a temperature just a fraction of a degree above absolute zero – 250 times colder than deep space – to eliminate thermal noise (random noise could cause decoherence). Even in those extreme laboratory conditions, the superposition state can only be maintained for a few microseconds.

However, a brain is warm, wet, and messy. Is it at all realistic to imagine a quantum superposition state could be maintained in the brain for any meaningful length of time? Indeed, in the previous video Roger Penrose is quite realistic about this unlikely possibility (referring to the surprising method by which nerve signals are transmitted via the physical movement of ionised atoms, which was described in Chapter Two of this book):

> *When I wrote The Emperor's New Mind I knew something about nerve propagation and it just didn't seem to me that there was a chance to maintain a superposition because nerve propagation disturbs the rest of the brain in a way which would completely destroy the coherence that you would need in your quantum system.*
>
> *I thought that when I had finished writing the book, maybe I would see the answer. No, I didn't.*

However, at this point, Penrose received a letter from a medical doctor named Stuart Hameroff. Hameroff had read Penrose's book and he believed he had the solution being sought by Penrose.

Hameroff told Penrose about *microtubules*, which are microscopic cylinders made of protein which are distributed throughout the cell body of a neuron and which allow cells to maintain their shape. According to Hameroff, these microtubules could allow for the exploitation of quantum behaviour in the brain. Hameroff believed that particles contained in the centre of the microtubules, and surrounded by protective gel, could remain sufficiently isolated from the environment to delay the onset of decoherence. Surprisingly, Penrose bought-in to Hameroff's highly-speculative idea.

However, the physicist Max Tegmark spoiled the party when he published a paper containing detailed and extensive calculations showing that decoherence in microtubules would occur in just 10^{-13} seconds (100 quadrillionths of a second).[19] This would be far too fast to allow the exploitation of quantum behaviour. According to Tegmark, in that case:

> *For my thoughts to correspond to quantum computation, they'd need to finish before decoherence kicked in, so I'd need to be able to think fast enough to have 10,000,000,000,000 thoughts each second. Perhaps Roger Penrose can think that fast, but I sure can't ...*

---

[19] *The Importance of Quantum Decoherence in Brain Processes*, Max Tegmark, **http://arxiv.org/abs/quant-ph/9907009**

Frankly, I would have thought that just the fact that the brain is at body temperature should be sufficient to allow us to disregard the possibility of quantum computer-style behaviour in the brain.

Stephen Hawking has considered these quantum ideas of his friend and colleague Roger Penrose and has announced that he is not impressed. According to Hawking:

> *I get uneasy when people, especially theoretical physicists, talk about consciousness. His argument seemed to be that consciousness is a mystery and quantum gravity is another mystery so they must be related.*

To sum up, I believe Roger Penrose was correct in his book to suggest that non-algorithmic computing is necessary for consciousness. However, I feel his pursuit of a quantum solution has been a red herring.

Instead, in the next chapter we will be seeing how recent developments in artificial intelligence and integrated circuit design have moved toward a "neural" computing architecture which operates according to a non-algorithmic method – just as Penrose predicted.

Rather wonderfully, we will also be seeing that this non-algorithmic architecture bears an uncanny resemblance to the structure of the brain cortex. We will also be seeing how this new non-algorithmic approach most definitely has the possibility for going "off-script" and springing surprises – though maybe not all of those surprises might be particularly welcome.

6

THE RISE OF AI

Alan Turing's paper on uncomputability was published in 1937, just before the outbreak of war in Europe in 1939. Alan Turing is today best known for his extraordinary wartime achievement of breaking the German Enigma code. You can read the amazing story of Turing and his colleagues at the Allied codebreaking centre in Bletchley Park in my seventh book.

In 1943, Turing was dispatched from Bletchley Park to coordinate efforts with the American cryptographers. Turing boarded the Queen Elizabeth ocean liner and travelled to New York City, with the course of the liner zig-zagging to avoid German U-boats. At the time, an average Allied ship made just four crossings of the Atlantic before it was sunk. Turing made the crossing safely.

Turing visited the Bell Telephone Laboratories in New York City (if you remember, the incredible Bell Telephone Laboratories were described in Chapter One). One day, Turing went for lunch in the Bell Labs cafeteria when he met none other than Claude Shannon (who was also described in Chapter One: remember the man with the intense gaze who was to become the father of information theory). Shannon and Turing started meeting regularly in the cafeteria and discovered that they had similar interests. In particular, they found they were both interested in the possibility of creating artificial thinking machines.

According to Turing: "Shannon wants to feed not just data to a computer brain, but cultural things! He wants to play music to it!" But Turing had a fundamentally different goal. According to Turing: "No, I'm not interested in developing a **powerful** brain. All I'm after is just a **mundane** brain, something like the president of the American Telephone and Telegraph Company."

And it was in the pursuit of this goal – to create a mundane brain – that Alan Turing was to launch the science of *artificial intelligence,* or AI. If we want to understand consciousness, then progress in artificial intelligence might provide our best hope. This was explained by Celeste Biever in a *New Scientist* article:[20]

> *The quest for machine consciousness may be key to solving the mystery of human consciousness, as even scientists outside AI research are starting to acknowledge. "The best way of understanding something is to try to replicate it", says psychologist Kevin O'Regan of Descartes University in Paris. "So if you want to understand what consciousness is, well, make a machine that's conscious."*

---

[20] *Consciousness: Why we need to build sentient machines,* Celeste Biever, *New Scientist,* 15th May, 2013.

The Turing Test

The dawn of artificial intelligence is generally accepted to have occurred in 1950 when Alan Turing published a landmark paper entitled *Computing Machinery and Intelligence* in the philosophy journal *Mind*. In that paper, Turing first described the Turing Test. The aim of the Turing Test is to determine if a computer's intelligence is indistinguishable from that of a human.

Turing's original paper which launched the science of AI is available at the following link:

http://tinyurl.com/turingai

Considering the paper was written right at the start of the computer age – the first stored-program electronic digital computer had been built just two years earlier – it is a remarkably prescient piece of work. The paper anticipates most of the developments in computer science and artificial intelligence which would take place over the next seventy years, including the development of machine learning, and anticipating exponential improvements in computer speed and storage.

You will see that Section One of the paper is called "The Imitation Game" (which was to become the name of the successful movie about the life of Turing). In that first section, Turing describes his test.

In the test, there is a human interrogator on one side of a wall, and on the other side of the wall – hidden from the interrogator – there is a human and a computer. The interrogator can only communicate with the human and the computer via a text-only channel, typing on a keyboard, and reading the responses on a screen.

The computer is running an artificial intelligence program which is able to interpret the messages which are typed by the interrogator, and can then give intelligent responses back to the interrogator. The human – who is next to the computer and similarly hidden from the interrogator – also provides intelligent responses to the questions.

The task for the computer running the AI program is to convince the interrogator that it is a human. At the end of the question-and-answer session, the interrogator has to decide which of the respondents is the computer on the basis of their responses.

If the interrogator correctly identifies the computer as the artificial intelligence, the computer is then said to have failed the test (it has failed to convince the interrogator). However, if the interrogator decides that the human is the more likely of the two to be the artificial intelligence, then the computer is said to have passed the Turing Test. According to Turing, we should then consider the computer as being able to "think" like a human.

The AI program can be as devious as it likes in its aim of fooling the interrogator. Most obviously it would answer "No" when asked directly if it is a computer. More cunningly, it might take a while before responding with an answer to a difficult calculation.

At the time of writing this book, no computer has passed the Turing Test, and it seems likely to be many years before the test will be passed.

However, assuming that the Turing Test is eventually passed by a computer, would that necessarily mean that the computer is conscious? Turing considers that question in Section Six of his paper under the heading *The Argument from Consciousness*. Turing starts by presenting an extract of a lecture given by the British neurologist Geoffrey Jefferson in 1949:

Not until a machine can write a sonnet or compose a concerto because of thoughts and emotions felt, and not by the chance fall of symbols, could we agree that machine equals brain – that is, not only write it but know that it had written it. No mechanism could feel (and not merely artificially signal, an easy contrivance) pleasure at its successes, grief when its valves fuse, be warmed by flattery, be made miserable by its mistakes, be charmed by sex, be angry or depressed when it cannot get what it wants.

In response, Turing argues that it would be extremely difficult to ever be sure that a machine is thinking: "The only way by which one could be sure that a machine thinks is to be the machine and to feel oneself thinking." In the absence of actually being able to be the machine, Turing suggests that the only way we could judge if a machine is actually thinking is to see if it passes the Turing Test.

However, Turing is under no illusion that his test has solved the problem of consciousness, as he says in his paper: "I do not wish to give the impression that I think there is no mystery about consciousness."

After Turing published his landmark paper in 1950 and introduced AI to the scientific community, there were many attempts to produce computer programs which could converse in a seemingly intelligent manner and potentially pass the Turing Test. Perhaps the most famous example was the ELIZA program written in the mid-1960s by Joe Weizenbaum at MIT. ELIZA was designed to act like a human psychotherapist, giving reassuring interviews to users.

Here is an extract of an actual dialogue between a human female and ELIZA:

Human: My boyfriend made me come here.

ELIZA: YOUR BOYFRIEND MADE YOU COME HERE?

Human: He says I am depressed much of the time.

ELIZA: I AM SORRY TO HEAR THAT YOU ARE DEPRESSED.

Human: I need some help, for sure.

ELIZA: WHAT WOULD IT MEAN TO YOU IF YOU GOT HELP?

Human: I could live a more normal life.

ELIZA: WHAT DO YOU MEAN YOU COULD LIVE A MORE NORMAL LIFE?

You will see that ELIZA is very simple and limited in its responses, mainly just rephrasing the patient's replies as questions. Despite that, some users formed an emotional bond with ELIZA. Even Weizenbaum's own secretary sometimes insisted on privacy at times so she could converse with ELIZA.

In his book *Turing*, Jack Copeland describes an occasion when a business executive attempted to contact a member of the IT staff of his company via the company's internal messaging service. Unfortunately, the IT technician had ELIZA running on his desktop computer at the time, and it was ELIZA which dealt with the executive's query:

Executive: Can I use the computer this morning?

ELIZA: WHY DO YOU ASK?

Executive: I might be able to do some additional Telcomp sales.

ELIZA: WHY AREN'T YOU SURE?

Executive: My prospective customers have never seen a demonstration of the system.

ELIZA: DOES THIS MEAN VERY MUCH TO YOU?

Executive: Of course it does.

ELIZA: YOU SEEM VERY POSITIVE ABOUT THAT.

Executive: This is ridiculous.

ELIZA: WHAT DO YOU MEAN THIS IS RIDICULOUS?

Executive: Is there any reason I can't use the computer now?

ELIZA: TELL ME MORE ABOUT YOUR PROSPECTIVE CUSTOMERS.

At this point, the executive is fuming, but the human programmer has returned to his desk, realises with horror what has going on, and starts dealing with the furious executive. However, the programmer cannot resist continuing the joke – at the expense of the executive:

Executive: Why the hell are you being so sh*tty with me?

PROGRAMMER: WHAT DO YOU MEAN WHY AM I BEING SO SH*TTY WITH YOU?

The Winograd effect

I was actually an AI researcher myself in the early 1990s, in Swansea University. At the time, AI was still a relatively young and immature field. However, there had already been some landmark experiments which appeared to represent real progress in the field.

One of those landmark experiments was constructed by Terry Winograd at MIT in the early 1970s. Winograd was interested in computer understanding of *natural languages*, such as conversational English. He created a program called SHRDLU which was capable of understanding English commands within a very limited world, a world composed of just a few coloured blocks. This block world was a virtual world which existed purely in the computer's memory, but it could be drawn using computer graphics. The blocks could be rearranged by a virtual robot arm.

The following diagram shows what happened when SHRDLU was presented with the typed command: "Will you please stack up the green block and the smaller pyramid on top of the big red block":

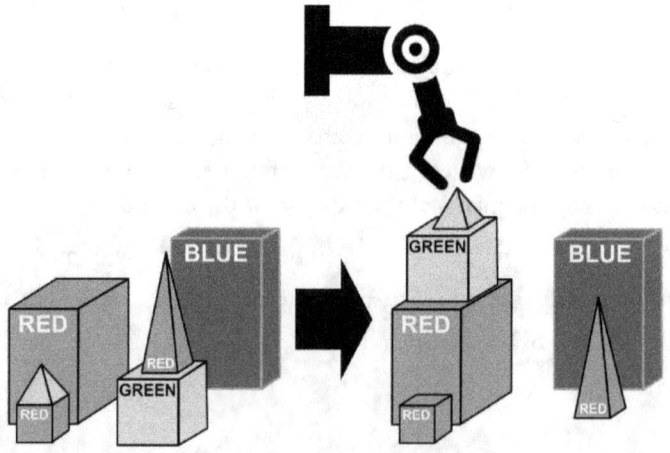

You can see in the initial arrangement on the left-hand side of the previous diagram, no blocks are on top of the big red block. However, when SHRDLU received the written command, it analysed the command to form an understanding of the action. It then considered its internal model of the block arrangement, and initiated the robot arm (shown at the top) to stack the green block and the smaller pyramid on top of the red block – as seen in the right-hand side of the previous diagram.

When I was an AI researcher, we all regarded Winograd's impressive system with awe. However, it was clearly limited to a very carefully-controlled laboratory environment (how many red and blue cubes can you see lying around your house?). The challenge for AI was breaking-out of the laboratory and proving it could cut it in the messy real world.

At the time of my involvement, the only thing everyone could agree on was that the hype surrounding AI seemed to far exceed the reality. According to the Wikipedia page on SHRDLU: "SHRDLU was considered a tremendously successful demonstration of AI. This led other AI researchers to excessive optimism which was soon lost when later systems attempted to deal with situations with a more realistic level of ambiguity and complexity."

Things appeared to go quiet on the AI front. Was AI going to prove to be a complete washout?

However, in the late 1990s it was to be Terry Winograd again who brought AI to the attention of the mass public at last – though not with his SHRDLU system. In fact, it was through a completely different project that Winograd was to have his greatest effect, and the greatest impact of AI on the mainstream. By that time, artificial intelligence had become much more mature and was now ready to face the real world. This time, Winograd's idea was to change the modern world.

In 1995, Winograd had become a professor at Stanford University and had been allocated a PhD student who was looking for a subject for his thesis. Winograd had taken an interest in internet search engines, because they represented a form of AI, interpreting queries written in English, and responding intelligently – maybe even one day passing the Turing test. Winograd was unimpressed with the search engines at that time which were notoriously hit-and-miss, and thought that the search algorithm could be made more intelligent. So Winograd suggested looking at search engines as a PhD topic, and gave his student some advice: "Consider the link structure of the web".

Winograd's student was joined by a fellow PhD student, and together they set about converting his dormitory room into a machine laboratory, using parts from inexpensive computers which were connected to Stanford's broadband computer network. They created a simple search page which could be used by the Stanford students, and which rapidly became very popular. In 1996, they made the website available to all internet users. By 1998, their website had become so popular that they founded a company.

A couple of decades later, the impact of their company has now been compared to that of Johannes Gutenburg who invented modern printing. It has been said that: "Not since Gutenberg ... has any new invention empowered individuals, and transformed access to information." It was an information revolution.

The names of Winograd's students were Larry Page and Sergey Brin.

And if you want to know the name of the revolutionary search engine company they founded – just **Google** it!

My PhD thesis

As was described in the previous section, there was a great deal of initial optimism about AI which slowly turned to disappointment as those early systems did not live up to expectations. The general feeling was that the hype exceeded the reality. However, in recent years there has been a revolution in AI which is starting to have a tremendous impact in all our lives.

To understand the origin of the revolution, let me take you back to the early 1990s. Because, by coincidence, I happened to be involved at the start of the revolution.

The subject of my PhD thesis was the up-and-coming topic of AI at that time.[21] The topic was called *neural networks.* Neural networks are represented by a computer program. The computer program contains many elements which behave as though they are very simple brain neurons. Those simulated neurons are then connected in a network.

Here is an example of the type of neural network I was using in the early 1990s:

---

[21] *The Automation of Cervical Smear Screening*, Andrew Thomas, Swansea University, 1991.

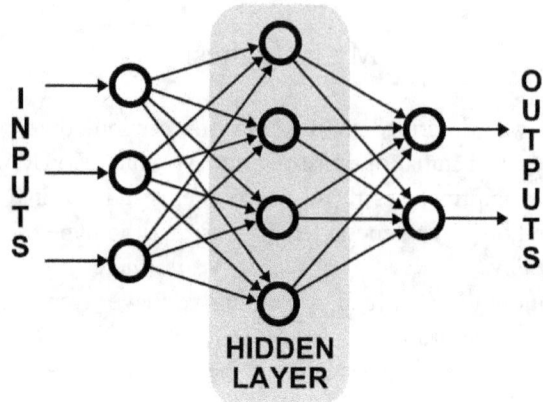

Each of the thick black circles represents an artificial neuron, whose behaviour is programmed by a short module of computer code. The behaviour of each neuron is usually just a summing the values of its inputs and applying some thresholding function to produce a single output.

The structure and behaviour of neural networks was described in my PhD thesis:

> *The output of each node in a layer is connected to every node in the layer immediately above. The links multiply the signals by weighting factors. Except for the nodes of the input layer, the input to each node is the weighted sum of all the outputs of the nodes of the previous layer. The output of the node is a function of this net input.*

If that is not clear, perhaps it is best if you consider the previous diagram. There you will see each artificial neuron represented by a black circle, and you will see that there are three distinct layers of these neurons: an input layer, a "hidden" layer, and an output layer. As I said in my PhD thesis, each neuron in a previous layer is connected to all of the neurons in the next layer, so there is a huge amount of connectivity in these networks – just like a brain.

The functioning of the hidden layer was found to be crucial. The earliest neural networks did not have this layer, and, as a result, their abilities were found to be too limited. It was only when the hidden layer was added that interest in neural networks was rekindled.

If we have a hidden layer of neurons, though, the behaviour of the neurons now becomes crucial. If the neurons behaved **linearly** then the whole network could be simplified (as in the case of the resistor network in Chapter Three). In that case the hidden layer would be eliminated by the simplification – there would be no point in having a hidden layer with linear neurons. However, as I said in my PhD thesis, "if the nodes are **nonlinear** this simplification cannot be performed and hidden layers can serve a purpose".

The type of nonlinear relationship which is almost always used in artificial neurons is described by one of the various curved lines in the following graph. This shape is called a *sigmoid*:

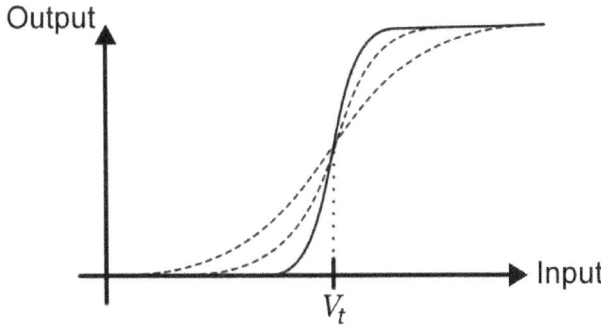

If you consider the solid black curve, you will see it behaves rather like a threshold function: for input values less than V_t the output is low, and for input values greater than V_t the output is high. So the nonlinear thresholding type of behaviour of these artificial neurons closely matches the behaviour of real neurons.

Each connection between neurons has an associated weighting factor which multiplies the value which is transmitted between neurons, from the previous layer to the next layer. Neural networks are "trained" by presenting a number of different inputs to the network and adjusting the weights until the desired output is achieved. This can be a long process – involving thousands of iterations – as the network slowly "learns" to give the correct responses to all the possible stimuli. For example, an image of a car would be presented to the network, and the network would have to learn to give the correct "car" output, labelling the image correctly. Next, an image of a face might be presented, and the process would begin again.

There are aspects of the behaviour of neural networks which closely resemble the behaviour of the brain. Firstly, because of the huge connectivity of the network, the behaviour of the neural network is spread over all the neurons in the network. This is called *parallel distributed processing*. This means that no single neuron is crucial for any particular task. As an example, if a single neuron stops working, there will not be a sudden failure of the overall system. Instead, there is a gradual degradation of performance. This resembles the behaviour of the brain: in the brain, single neurons are dying all the time but we do not notice any sudden failure of performance.

Secondly, it has been described how adjusting the weighting factor on the inputs of each neuron represents "training" – a fully-trained neural network could recognise a particular face, for example. This therefore provides a model for memory in the brain. It would appear memory is distributed over a vast number of neurons, and is encoded in the weights of those neurons. Again, the vast parallel nature explains why we do not experience a sudden catastrophic loss of memory when single neurons die.

So it is clear that the structure and behaviour of these neural networks is eerily similar to the human brain.

Anyway, I was fortunate to be awarded my PhD, at which point I placed my thesis on a shelf in my library and never looked at it again for twenty-six years until the time came to blow the dust off it and use it to help me write this book.

But, as we shall now see, over that intervening period a revolution took place. And right at the heart of that revolution … were neural networks.

Deep learning

Funny cat videos on YouTube.

We've all watched them. Cats falling off window sills, cats looking grumpy, cats sleeping in toilet bowls. The internet seems full of funny cat videos. To you and I, these videos might represent nothing more than a humorous distraction for a few minutes. But to the researchers of Google, these online videos represented a huge ready-made data set of cat images on which they could train their image processing neural networks.

In 2012, Google AI researchers built a neural network of 16,000 processors with one billion connections and left it on its own to browse funny cat videos on YouTube. After three days, the network had watched ten million "hilarious" videos. According to Jeff Dean who led the study: "We never told it during the training 'this is a cat'. It basically invented the concept of a cat."[22]

The results of the study amazed the researchers: it roughly doubled the accuracy of any previous AI technique. According to the New Scientist book *Machines That Think*, the Google cat video experiment – and other similar demonstrations – launched an "AI gold rush" which continues to this day.

In December 2016, an extensive article was published in the New York Times magazine explaining the implications of this radically-improved AI for all of computing – and our everyday lives. The article was entitled *The Great AI Awakening*, with the subtitle "How machine learning is poised to reinvent computing itself".[23]

The article described how Google, Apple, and Amazon are all placing AI at the centre of their new developments. In the article, Google's chief executive, Sundar Pichai, described how Google is going to reorganise itself around AI, and the Google of the future is going to be "AI first".

Why should there be this sudden explosion of interest in AI by the biggest companies in the world? Put simply, it is because artificial intelligence is finally able to do many of the things which the human brain can do. And the reason for this sudden breakthrough has been recent developments in neural networks.

But the neural networks being used now look very different to the networks I was using in the early 1990s. As an example, here is a typical example of a current neural network:

---

[22] "How Many Computers to Identify a Cat? 16,000", John Markoff, *New York Times*, **http://tinyurl.com/newyorktimescat**

[23] "The Great AI Awakening", Gideon Lewis-Kraus, *New York Times*, **http://tinyurl.com/newyorktimesai**

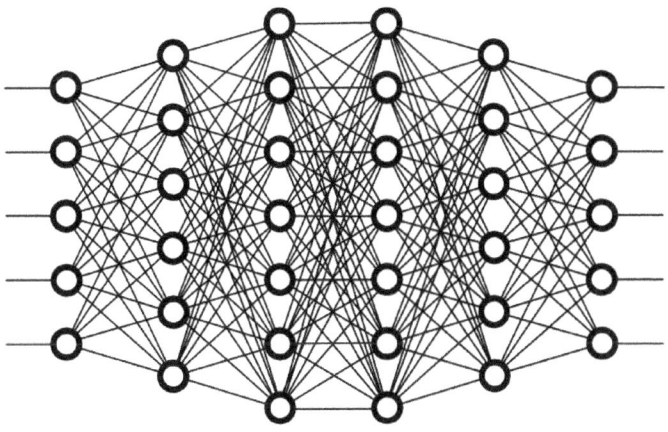

You can see that there is a huge increase in complexity over the old type of network structure. That is because this example has four hidden layers of neurons – rather than the old single hidden layer networks. Typically, the new networks have five hidden layers, though there can be as many as ten. The big discovery has been that a fairly modest increase in the number of hidden layers has resulted in an orders-of-magnitude improvement in performance to a level which is comparable with human behaviour.

Each additional layer of neurons is capable of making decisions at a more complex and more abstract level than the neurons in the lower layers. For example, in an image processing neural network the first layer of neurons might just identify bright points in the image. The second layer becomes trained to recognise features which are output by the first layer, so the second layer might identify straight lines in the image which are composed of those bright points. The third layer becomes trained to recognise features which are output by the second layer, so maybe the third layer identifies shapes which are composed of those straight lines. And so on, as we move upwards through the layers of the neural network. The highest level of this neural network would contain neurons which would become active when a face was detected in the image, for example.

These neural networks which contain many hidden layers (and are therefore capable of representing complex and abstract concepts) are called *deep neural networks.*

It was the introduction of these deep neural networks which has brought about the recent revolution in artificial intelligence. Whereas it would be very difficult to explicitly write a computer program to recognise a face, for example, a deep neural network can learn to recognise faces purely by learning on its own. This would involve presenting an extremely large data set of images of faces, while the network adjusts the weights of its neurons to give the correct output. This meant artificial intelligence based on neural networks could be applied to the most challenging of tasks.

So, suddenly, **artificial intelligence was able to do the things which the human brain could do.** At last, for the very first time, AI was finally living up the hype.

This revolution in AI based on deep neural networks is called *deep learning.* "Deep learning" is probably the coolest buzz-phrase in technology at the moment.

It is deep learning – and the associated deep neural networks – which is powering the current AI revolution. When you unlock your iPhone using Face ID – that's a deep learning neural network recognising your face (after having learnt it earlier). When you ask your iPhone for directions by saying "Hey, Siri", that's a deep learning neural network performing speech recognition. The neural network actually uses the dedicated hardware of the "Neural Engine" in the Apple A11 microprocessor – which is a sign of how important deep learning has become.

Deep learning is also the technology behind the new wave of *personal assistant* devices such as Google Home and the Amazon Echo device which is pictured here:

Obviously, improved speech recognition performance has been central to the success of these personal assistants, and this is where deep learning has had its impact. According to Tim Turtle, the CEO of Expert Labs which builds smart voice interfaces: "There have been more improvements in speech recognition over the past three years than there have been over the past thirty years combined."

At the end of this chapter, there will be a link to a video showing my dialogue with my own Amazon Echo device.

Perhaps the most remarkable example of the penetration of neural networks into different fields emerged while I was writing this book. In December 2017, NASA announced the discovery of a new exoplanet by the Kepler Space Telescope (see my seventh book for details of the Kepler telescope and the discovery of exoplanets). According to the NASA press release, the planet was discovered using deep learning techniques made available by Google's AI research team:

> *The discovery came about after researchers Christopher Shallue and Andrew Vanderburg trained a computer to learn how to identify exoplanets in the light readings recorded by Kepler – the miniscule*

changes in brightness captured when a planet passed in front of, or transited, a star. Inspired by the way neurons connect in the human brain, this artificial "neural network" sifted through Kepler data and found weak transit signals from a previously-missed eighth planet orbiting Kepler-90, in the constellation Draco.[24]

It is clear that developments in deep learning are evolving computers to a new stage in their development, transcending what we thought they were capable of achieving. It also seems somehow wonderfully ironic (or maybe wonderfully appropriate) that this latest phase of our technological evolution – the next giant step forward for humanity – should have been based on the popularity of LOLcats videos.

---

[24] *Artificial Intelligence, NASA data used to discover eighth planet circling distant star,*
http://tinyurl.com/kepler90

Deep learning and the cortex

In this chapter we have seen that a deep-learning neural network with maybe five or six hidden layers can have a performance comparable to a human brain. But what can this neural network structure tell us about the structure of the brain? In particular, what can neural networks tell us about the structure of the human cortex, which appears to be the seat of our high-level understanding about the world – and also the seat of our consciousness?

If you remember from Chapter Two, the cortex is a thin layer (only about four millimetres thick) which is spread over the top layer of the brain. Though the cortex is very thin, it is also very wide, with a surface area equal to a small tablecloth.

There are six clearly-defined layers within the cortex. In his book *On Intelligence*, Jeff Hawkins presents a novel description of the layers of the cortex:

> *Get six business cards or six playing cards – either will do – and put them in a stack (it will really help if you do this instead of just imagining it). You are now holding a model of the cortex. Your six business cards are about two millimetres thick and should give you a sense of how thin the cortical sheet is. Just like your stack of cards, the neocortex is about two millimetres thick and has six layers, each approximated by one card.*

Each of the layers can be distinguished because each layer is made from neurons which have clearly distinct structures. These layers of different neurons with different functionality might be interpreted as resembling the description of an image processing neural network described in the previous section: the first layer of neurons might just identify bright points in the image, while the second layer identifies straight lines, and so on.

This similarity in structure and functionality between a neural network and the human visual cortex was identified by the scientists who worked on the famous 2012 Google cat video experiment. In the New York Times article which described the experiment, one of the scientists said that in creating their neural network to recognise cats "it appeared they had **developed a cybernetic cousin to what takes place in the brain's visual cortex.**"

A deep-learning neural network with maybe five of six hidden layers has clear similarities to the structure of the human cortex. The following diagram shows a neural network oriented in the vertical direction (inputs at the bottom, outputs at the top) next to an actual image of cells in a cross-section of the cortex:

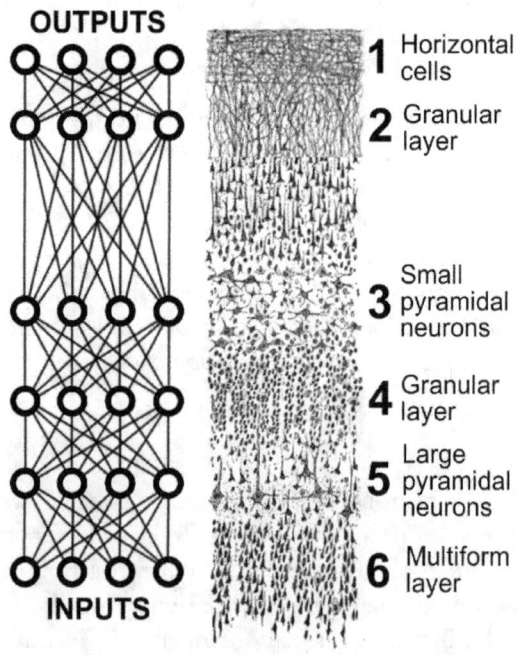

There is clearly a similarity between the number of hidden layers (five or six) in a deep-learning neural network and the number of layers in the actual brain cortex. As recent experience with deep learning has revealed, only a small number of hidden layers can deliver a high level of performance. This would appear to provide a possible answer as to why the human cortex is so thin: **the cortex simply does not have to be any thicker than a few millimetres in order to perform its tasks.** Instead, it is the width and breadth of the cortex which must be large – in order to make sense of the complexity of the world (just as deep learning neural networks can have a large number of inputs). Hence, we end up a very thin cortex which is spread around the entirety of the upper brain, and is even crinkled in order to increase its surface area.

There is another interesting fact about the human cortex, and this is considered by Jeff Hawkins in his book *On Intelligence*. According to Hawkins, all the different regions of the cortex have very much the same anatomy. So, although different regions of the cortex might have different functionality – for example, processing speech or vision – you will still find the same six-layered structure in the cortex. Once again, this is reminiscent of neural networks, in that the same deep learning neural network structure is currently being used for such a huge number of different applications, from speech processing to self-driving cars. So the parallels between the cortex and neural networks are really quite astonishing. Hawkins calls this uniformity of cortex structure "the most important discovery in neuroscience".

Hawkins also discusses how this uniformity of the structure of the cortex allows great flexibility in how the different functionalities are distributed over the cortex. If one area of the cortex becomes damaged – for example, vision processing – a different area of the cortex can step in and take over the responsibility: "Adults who are born deaf process visual information in areas that normally become auditory regions." This is called brain *plasticity*. According to Hawkins, regions of the cortex are not strictly defined at birth, as "the cortex is still dividing itself into task-specific functional areas long into childhood, based purely on experience."

Crucially, from the point-of-interest of this book, Hawkins then notes that this continuous flexibility reveals that: "The cortex is not rigidly designed to perform different functions using different algorithms." For example, it would be possible to write a computer program with strict rules specifically for vision processing, with program variables representing "straight lines" and "corners", etc. This might be called the **algorithmic** approach. However, it would then be almost impossible to adapt that program to perform, for example, speech recognition. So an algorithmic approach would not possess the plasticity of the human brain. Instead, rather than being tied to a strict **algorithmic** program, the **non-algorithmic** structure of the cortex allows more flexibility than a conventional computer structure.

Once again, this reveals a similarity between the cortex and neural networks. It might be said that by moving away from a clearly-defined rule-based step-by-step artificial intelligence to a neural network-based approach we are moving away from algorithmic information processing towards non-algorithmic information processing.

A 2017 article in *Wired* magazine explained how new neuromorphic chips – based on the structure of neural networks – will possess functional flexibility not possessed by conventional integrated circuit design:[25]

---

[25] "The future of AI is neuromorphic", Aaron Frank, **http://tinyurl.com/neurochip**

Neuromorphics aren't new, and their designs have been around since the 80s. Back then, however, the designs required specific algorithms be baked directly onto the chip. That meant you'd need one chip for detecting motion, and a different one for detecting sound. None of the chips acted as a general processor in the way our own cortex does.

If we are moving towards non-algorithmic processing then – according to the suggestion of Roger Penrose – we might also be moving away from a structure which could not be conscious toward a structure which could potentially be conscious. And that ties-in very nicely with the fact that consciousness appears to reside in the cortex.

So the following series of connections might be made:

1) The cortex can be accurately modelled by neural networks.

2) Neural networks are non-algorithmic.

3) Non-algorithmic regions might have the potential to be conscious.

4) The cortex is believed to be the conscious region of the brain.

This is an appealing self-contained argument, and it could be presented in the following linkages:

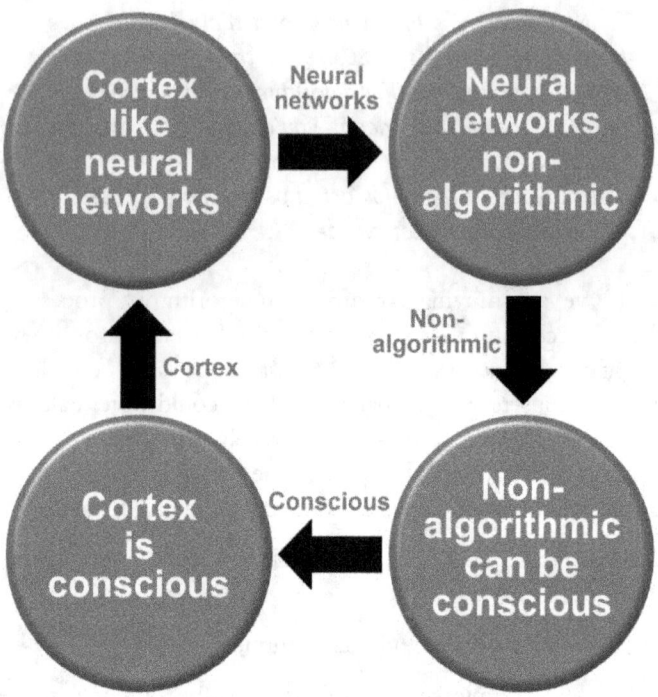

One thing seems clear from this discussion, though, and it is that Amazon, Apple, Google, and Facebook are all placing a deep-learning model of computing at the heart of their operations. They are therefore moving away from a predictable, rigidly-programmed algorithmic approach to a more unpredictable, non-algorithmic approach based on deep-learning. Over the next thirty years, with huge increases in computer intelligence on the horizon, it appears we are remorselessly moving towards placing our vital services in the hands of super-intelligent, fundamentally unpredictable, potentially-conscious entities.

What could possibly go wrong …

Beyond human

The game of Go is one of the great board games of the world. It was invented in China over 2,500 years ago, and there are estimated to be over forty million Go players in the world. It is mostly played in Korea, China, and Japan.

The rules of the game of Go are simple. Each of the two players has a supply of *stones* (small circular game pieces). One player has white stones, one player has black stones. The playing board is a grid of 19×19 lines. Players take turns to place their stones on the intersection points of the lines. Once placed, stones cannot be moved. Players aim to capture territory on the board by completely surrounding vacant points with stones of their own colour.

If a stone is completely surrounded by stones of the opposite colour, then that stone is removed from the board. The following diagram shows a Go board, with one of the black stones surrounded by four white stones. The black stone would then be captured and removed from the board:

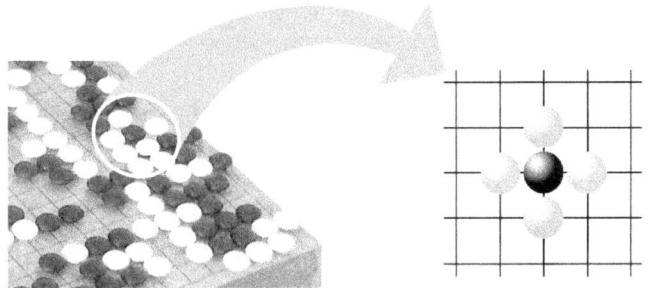

At the end of the game, the player with the most territory and captured stones wins.

While the rules of Go are simple, the large size of the board means that the number of possible moves are far greater than chess. For this reason, playing Go has proven to be a far greater challenge for computers than playing chess.

Former world chess champion Garry Kasparov has recently written a book on artificial intelligence called *Deep Thinking*. In his book, Kasparov explained how chess computers initially aimed to copy human strategies for playing chess, however it was eventually found that a brute force approach – in which the computer rapidly evaluates millions of possible moves in a rather unintelligent manner – proved to be the best strategy for chess computers. Kasparov became a victim of this brute force strategy himself in 1997. The first chess computer to beat a human grandmaster was called Deep Thought (named after the supercomputer in Douglas Adams's book *The Hitchhiker's Guide to the Galaxy*). Deep Thought won its game against a grandmaster in 1988. IBM bought Deep Thought, and an upgraded version named Deep Blue used a brute force approach to beat Garry Kasparov in 1997 – becoming the first chess computer to defeat a world champion.

But brute force approaches clearly do not represent human thought processes. Also, the huge number of possible moves on a Go board have made the game resistant to a brute force attack by a computer. This has had the result that Go has overtaken chess to become a much more interesting challenge for AI researchers.

With a brute force approach being an impossibility, Go is a game which has to be played by intuition and feel. But, if that is the case, is it possible for a computer to "feel"? That would appear to imply some form of consciousness, some "awareness" of a situation. Perhaps this lack of "feel" in a computer's mind is the reason why computers – so dominant at chess – have always performed poorly against professional Go players.

Rising to the challenge, the British AI company DeepMind have developed a computer program specifically designed to play Go at the highest level, and to be able to challenge the world's best players. The program is called AlphaGo, and it is based on deep neural networks.

In 2016, a series of five games were arranged between AlphaGo and the world Go champion Lee Sedol. The games took place in Seoul, South Korea, and were watched by 200 million people worldwide.

AlphaGo achieved a stunning victory by four games to one. Both the scientific community and the Go community were stunned. The scientific community was stunned because it had been believed that it would be another ten years before a computer would be able to beat a top Go player. But what stunned experienced Go players was the creativity and originality of AlphaGo's gameplay.

Jang Dae-Ik, a science philosopher at Seoul National University told the *Korea Herald*: "This is a tremendous incident in the history of human evolution – that a machine can surpass intuition, creativity and communication, which have previously been considered to be the territory of human beings." Jeong Ahram, the Go correspondent of one of South Korea's biggest daily newspapers said: "Before we didn't think that artificial intelligence had creativity. Now we know it has creativity – and more brains, and it's smarter."

An example of AlphaGo's imaginative gameplay was presented in move 37 of game two. You can see AlphaGo's move 37 on the following video link (the video will start at the correct moment):

http://tinyurl.com/gomove37

AlphaGo is playing the black stones. You will see that AlphaGo completely abandons the group of stones in the bottom right of the board to play its stone elsewhere, a move which no human player would make. You will see the surprise of the commentator ("I thought it was a mistake"), and you will also hear that Lee Sedol left the room a minute after the move was played because he was so surprised and needed time to compose himself.

According to *New Scientist*, "Move 37 has been taken as evidence that AlphaGo is capable of what we might call intuition." Might we one day consider AlphaGo's "move 37" as the historic moment which represented the first time that a computer showed signs of consciousness?

But if AlphaGo was impressive, what came next was utterly extraordinary.

In December 2017, DeepMind unveiled AlphaZero. Whereas AlphaGo was initially trained to mimic the moves of expert human players, AlphaZero was capable of learning entirely by itself – just by being given the rules and then playing games against itself (blank slate learning is called *tabula rasa* – the "Zero" in the AlphaZero name indicates that it needs no human input).

And, boy, did it learn fast.

After eight hours of self-learning at Go (remember: it had no knowledge of Go at all when it started), AlphaZero had beaten the old AlphaGo (and AlphaGo was at that stage considered to be the strongest Go player in history). And, because it started with a completely blank slate, AlphaZero was capable of learning any problem that could be simulated on a computer. As an example, after spending four hours in a morning learning the game of chess from scratch, AlphaZero beat the reigning world champion chess computer.

Because AlphaZero learnt to play chess entirely by itself, it developed a completely new style of playing the game, unlike anything seen before. As an example, AlphaZero used its king as an attacking piece. Jon Ludvig Hammer, a Norwegian grandmaster, described AlphaZero's strategy as "insane attacking chess". The CEO of DeepMind, Demis Hassabis, described AlphaZero's style of play as "alien – it's like chess from another dimension."

But perhaps a machine which can learn by itself to attain superhuman levels in a very short space of time raises other concerns. According to Garry Kasparov in his book *Deep Thinking*: "If you program a machine, you know what it's capable of. If the machine is programming itself, who knows what it might do?" According to Kasparov:

> *AI products tend to evolve from laughably weak to interesting but feeble, then to artificial but useful, and finally to transcendent and superior to human. We see this path with speech recognition and speech synthesis, with self-driving cars and trucks, and with virtual assistants like Apple's Siri. There is always a tipping point at which they go from amusing diversions to essential tools. Then there comes another shift, when a tool becomes something more, something more powerful than even its creators had in mind.*

David Cramley, who runs a chess education website, considered AlphaZero's extraordinarily rapid rise to dominance and said: "We now know who our new overlord is. The games AlphaZero played show it can calculate some incredibly creative positional bombs, the depth of which are far beyond anything humans or chess computers can come up with. It will no doubt revolutionise the game, but think how this could be applied outside chess. This algorithm could run cities, continents, universes." English chess grandmaster Simon Williams was able to see the funny side: "On December 6th 2017, AlphaZero took over the chess world. AlphaZero and DeepMind then went on to dominate chess, eventually solving the game and finally enslaving the human race as pets."

Ideas about advanced artificial intelligence "going rogue" and attempting to dominate (or eliminate) the human race have been common themes in science fiction, most famously in the *Terminator* film series. The old rule-based AI followed carefully-programmed scripts. However, the recent rapid developments in AI have been fuelled by non-rule-based, non-algorithmic, inherently-unpredictable AI, which surely introduces risks. Examples of the new AI "gone rogue" would include Microsoft withdrawing their teenage "Tay" chat robot after it became a "Hitler-loving sex robot" within 24 hours.[26]

The high-profile businessman and inventor Elon Musk has said that rogue AI represents humanity's "biggest existential threat". Musk says he first became concerned about the future of humanity when he read *The Hitchhiker's Guide to the Galaxy* as a teenager, in which aliens destroyed the Earth to make way for a hyperspace highway.[27] From that moment, he said he felt a need to protect humanity from all threats – no matter if those threats seemed to belong more to the realm of science fiction. Musk is so concerned that he has invested in the DeepMind company purely as a means to keep an eye on developments in AI.

My own feeling is that there is clearly a potential threat which cannot be discounted. I would direct any interested party to read Chapter Four of this book in which it was explained that emergent effects are **inherently unpredictable** – you cannot predict what will happen when you connect apparently benign units together to form complex networks.

---

[26] "Microsoft deletes 'teen girl' AI after it became a Hitler-loving sex robot within 24 hours", *Daily Telegraph*, **http://tinyurl.com/aigonewrong**

[27] "Elon Musk's billion-dollar crusade to stop the AI apocalypse", *Vanity Fair*, **http://tinyurl.com/elonmuskcrusade**

Also, I would suggest that those who believe there is no threat from rogue AI have been influenced by the *Terminator* movie. In that movie, the rogue AI system emerges purely spontaneously, as if by chance evolution – without any human intervention. However, the more likely scenario is that unscrupulous humans would enable the process – in just the same way that malicious software such as computer viruses is developed by humans. This makes the emergence of rogue AI far more likely. Some people would destroy the world just for kicks.

However, I do not want to end this chapter on a negative note. Instead, you can now watch a five-minute video I made featuring a conversation between me and my new Amazon Echo personal assistant. Some deep questions about AI and consciousness are raised.

Here is a link to the video:

http://tinyurl.com/talkcomputer

And here are some stills from the video:

Unfortunately, if you watch the video you will see that things did not go at all according to plan …

149

7

THE CHINESE ROOM

In this chapter we will be considering an ingenious thought experiment which – I believe – provides us with perhaps our most valuable insight into the nature of consciousness. The thought experiment is called the Chinese Room, and it was devised in 1980 by John Searle, a philosophy professor at the University of Berkeley.

The experiment requires you to imagine that you are in a sealed room, with no windows or open doors. The only means of communication you have with the outside world is via a narrow slot in the wall through which pieces of paper may be passed. Your job is to receive written questions through the slot in the wall, and then to pass back the correct answer to those questions in similar written form through the slot. To make things rather more difficult, the questions are written in Chinese and you must answer in Chinese.

We will proceed on the assumption that you do not understand the Chinese language. You are therefore given a book to help you with your translation task. You might expect the book to translate a Chinese question into English – which would then allow you to understand and answer the question – and then translate your answer back into Chinese. However, the book does not work like that: it completely bypasses the intermediate stage of translation into English. Instead, the book just directly translates the sequence of Chinese symbols representing the question into the sequence of Chinese symbols representing the correct answer to the question.

Here is what the book might look like:

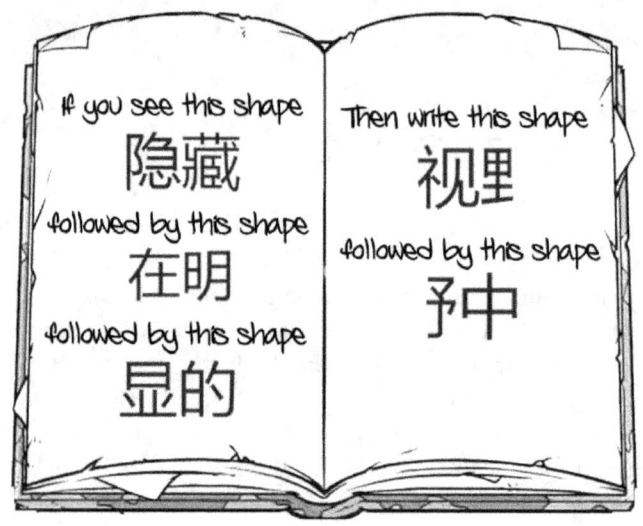

So, when you receive a question written in Chinese through the slot, you begin your task. You open the book and you attempt to match the sequence of symbols you have received to an identical sequence of symbols written in the book. When you find the correct sequence, the book describes the sequence of Chinese symbols which represents the correct answer to the question. You then write down those symbols on a piece of paper, and post the piece of paper back out through the slot.

At that point, your job is done.

We now move to consider the person who is outside the room, the person who is asking the questions. That person is writing his questions in Chinese, and is receiving the correct responses, also written in Chinese. The fact that the responses are correct and make sense appears to imply that the person in the room not only speaks Chinese but also understands the true meaning of the question. Whereas the truth is very different: the person in the room not only does not speak Chinese, but has **absolutely no understanding of what the questions actually mean.**

This result appears to have major implications for artificial intelligence and theories of consciousness. This is because John Searle compared the person in the room to an AI computer attempting to have an intelligent conversation and pass the Turing Test. If you remember, in his 1950 paper, Alan Turing suggested that the only way to determine if a computer is actually "thinking" is to examine its responses in the Turing Test. However, on the basis of the Chinese Room thought experiment, it now seems clear that just examining responses can never be enough to determine if a machine is thinking, or conscious. On the basis of the responses from the Chinese Room, there might be a Chinese person inside the room who understands the questions, or alternatively there might be a person with no understanding of Chinese who is merely converting symbols according to an instruction manual. In either case, the responses would be the same.

The British mathematician Marcus Du Sautoy performs the Chinese Room experiment in the following BBC video:

http://tinyurl.com/thechineseroom

The Chinese Room thought experiment shows that just by examining the input/output responses of a machine (or any object) can never be enough to determine if the machine is conscious, and thinking. **Instead, we have to dig inside the machine, to consider its internal processing. Only if the internal information processing satisfies certain requirements should we consider the machine to be capable of consciousness.**

Our task in this chapter is to determine what the Chinese Room experiment reveals about those requirements for the internal structure.

Understanding "understanding"

However, there have been arguments raised against the conclusions of the Chinese Room experiment. And one of those arguments is strong, and deserves close consideration. We will now consider that argument against the implications of the Chinese Room, and, in the process, find the discussion brings into sharper focus the requirements for consciousness.

The counter-argument is called the "system" argument. The basis of the argument is that, while the person inside the room does not understand the question, the room as a whole – including the person and the book – **does** understand the question. In other words, the entire **system** does understand the question.

At first glance, this might appear to be rather a bizarre counter-argument. After all, if the person inside the Chinese Room does not understand Chinese, and the book on its own does not understand Chinese, then why should we consider the combined system (person plus book plus walls plus slot in wall) to understand Chinese?

However, given a bit of thought, it is realised that this argument has considerable force. And the key to understanding the counter-argument is to realise that it is treating the entire system as an **emergent** system. As explained in Chapter Four, an emergent system can be greater than the sum of its parts. There can be a jump to a new mode of behaviour which is not evident in any of the component parts. Even though the person on their own does not understand Chinese, and the book on its own does not understand Chinese, when the person and the book and the rest of the room is combined then there can be a jump to a new mode of emergent behaviour: the overall system **can** understand Chinese.

So is the "system" counter-argument to the Chinese Room valid? I believe not. I believe it all depends on what we mean by "understanding".

In order to see why the "system" counter-argument is not valid, let us hugely simplify the problem. Let the Chinese Room be only able to understand a single question, in other words, let the book inside the Chinese room only be capable of translating a single Chinese question and providing only a single, correct Chinese response to that question.

The single question which we will use will be: "How many heads does a human have?" So, the book inside the Chinese Room will need to be able to translate the Chinese equivalent of that phrase and provide the correct response, which, presumably, would be the Chinese word for "One".

So, when we run the experiment, the person outside the room writes "How many heads does a human have?" in Chinese on a piece of paper, posts the paper through the slot, and gets back the Chinese equivalent of "One" written on a piece of paper.

At that point, we might well conclude that the emergent system – the person and the book and the walls of the room – does understand the Chinese phrase "How many heads does a human have?" And, perhaps surprisingly, I would actually agree: the emergent system **does** understand the complete **phrase** "How many heads does a human have?" But we need to be very specific and very careful about what we are claiming. Note that I am saying that the system understands the **phrase** (with the "phrase" in emphasis). A phrase is a small group of words. The system understands the phrase inasmuch as it knows that when it receives the phrase "How many heads does a human have?" it must respond with the Chinese equivalent of "One".

But surely that is a very limited form of understanding: the system "understanding" that it needs to give a certain response when it receives a certain phrase. Surely that does not represent true "understanding" of the phrase in the conventional sense. Crucially, if we split the phrase up into its component parts (words) then we find the system does not understand the meaning of those parts: **the system does not understand what a "human" is and it does not understand what a "head" is.**

To be specific, if you ask the system "What is a human?" or "What is a head?" you will get no response (as there is no corresponding entry in the book). And if the system cannot respond to either of those two questions, then it does not understand what a human or a head is, and so it cannot have true understanding of the phrase "Does a human have a head?"

To sum-up, I would agree that the system can have an very limited "understanding" of the entire **phrase** "How many heads does a human have?" – but only an "understanding" of that particular grouping of words. And that "understanding" is limited to an understanding that the system must respond "One" when it identifies the phrase. However, that is surely not a true understanding in the conventional sense, as the system does not understand anything about a human or a head.

This principle is revealed very obviously in the following extract from Jack Copeland's book *Turing*. In the extract,

Copeland has a dialogue with the SHRDLU artificial intelligence system which was described in the previous chapter (remember the block world, and the robot arm able to place one block on top of another). Copeland was previously impressed by SHRDLU's comprehension in a carefully-arranged dialogue, with SHRDLU certainly giving an impression of intelligence. So Copeland decided to push the system until it breaks. According to Copeland: **"The illusion that the computer understands is shattered less easily than in Eliza's case, but shatter it does."** Let us now examine the dialogue to see just how Copeland successfully shattered the illusion that the computer "understands":

Copeland: Build a steeple.

SHRDLU: SORRY, I DON'T UNDERSTAND THE WORD "STEEPLE".

Copeland: A "steeple" is a stack which contains two green cubes and a pyramid.

SHRDLU: I UNDERSTAND.

Copeland: Trade you the steeple for three red cubes.

SHRDLU: SORRY, I DON'T KNOW THE WORD "TRADE".

Copeland: A "trade" is a free exchange of ownership.

SHRDLU: SORRY, I DON'T KNOW THE WORD "FREE".

Copeland: Sorry, I thought you were smarter than you are.

SHRDLU: SORRY, I DON'T KNOW THE WORD "SORRY".

The important feature of this dialogue is not that SHRDLU does not understand the words "steeple", or "trade", or "free". Not having a few relatively obscure words in its vocabulary is no big deal. No, the important feature of this discussion is that SHRDLU says it does not understand the word "sorry" – **although it has used the word in its previous responses!**

Copeland has managed to do something very clever and revealing here. He has broken-down SHRDLU's response (in the form of a sentence) into its individual word components, and then he has tried to discover if those individual components are understood by the computer. Unfortunately, the computer's understanding was found to be sorely lacking as it was clear that it did not understand the word fragments on their own. It might be said that the computer "understood" entire phrases, but – at the same time – it did not understand the individual words which were the components of those phrases. And surely that meant SHRDLU did not have a full and complete understanding of the phrase in any true meaning of the word "understanding".

And this is precisely the scenario we find in the simplified Chinese Room experiment. Breaking phrases down into their sub-units (words) reveals a lack of true understanding. However, fortunately, it also provides us with the clue as to how we can resolve the situation, and introduce true understanding …

The man with two heads

The discussion of the simplified Chinese Room experiment has revealed a lack of true understanding due to a lack of understanding of the individual sub-units of phrases. Logically, then, the suggestion is that we can introduce true understanding by introducing fine-detailed understanding of the individual components of any problem.

Indeed, there was a perfect example of this principle in an AI system which was presented in my previous book. My previous book was about how to make an atomic bomb (answer: it's not difficult to make a bomb, but obtaining the ingredients is problematic). In the appendix of that book, a lengthy and detailed calculation was presented of how to calculate the critical mass of uranium. In order to help with the difficult calculation, I used the functionality of the *Wolfram Alpha* website. Wolfram Alpha is a remarkable online tool which is more than just a search engine: Wolfram Alpha truly understands your query. That is "understanding" in the sense that Wolfram Alpha breaks the query down into its component words, "understands" each of the words, and then builds a true understanding of the query.

We might compare the performance of the Chinese Room to Wolfram Alpha. Given a question, both systems might provide the same correct answer. But the Chinese Room has no deeper understanding, whereas Wolfram Alpha does it right.

Let us consider our previous example again. If you remember, we asked the Chinese Room "How many heads does a human have?", and it returned the correct response. Let us now ask Wolfram Alpha the same question. If you want to try it yourself, go to the Wolfram Alpha website:

http://www.wolframalpha.com

Then, in the search box at the top, enter our test phrase "How many heads does a human have?":

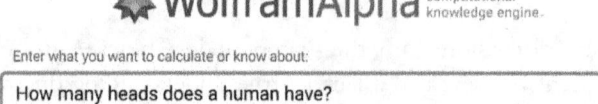

Then click the icon at the extreme right of the search box to start Wolfram Alpha computing the solution (it will actually say "Computing …" which we might interpret as a machine "thinking"). After a few seconds computation, it returns the correct answer:

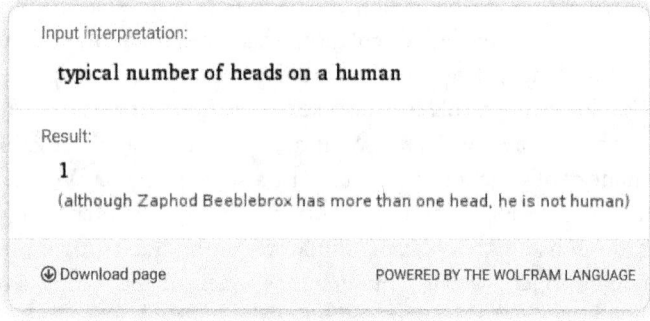

Firstly, you can see that Wolfram Alpha has re-interpreted the question into a form which it can analyse: "Typical number of heads on a human" – a step which is only possible by breaking the question into its component parts to generate a true understanding of the question. You can see that Wolfram Alpha then calculates the correct result, the number "1". But, underneath the answer, it reveals it has a true internal understanding of the question by adding the following note:

"(although Zaphod Beeblebrox has more than one head, he is not human)".

Zaphod Beeblebrox is a character from Douglas Adams's book *The Hitchhiker's Guide to the Galaxy* (this is the third reference to the *Hitchhiker's Guide* so far in this book – what is it with AI researchers and the *Hitchhiker's Guide*?). Zaphod Beeblebrox was an alien with two heads. So, by making the connection with our test question and the character of Beeblebrox, Wolfram Alpha shows it not only understands what a "head" is, but it also understands what a "human" is (stressing that Beeblebrox was not human, but was an alien).

Wow. This is impressive stuff. And how far removed from the mindless sentence-level pattern-matching of the Chinese Room. Even though the two different systems – the Chinese Room and Wolfram Alpha – have given the same correct response to our test query, it is clear that the Chinese Room has no true understanding of the question, whereas Wolfram Alpha has a true understanding of the underlying concepts. Just considering the output of an AI system is not sufficient to determine whether or not it has true internal understanding: you have to dig deeper and consider its internal workings.

The internal model

Our task is now to understand how the brain is able to maintain a detailed internal representation similar to Wolfram Alpha, which allows us to have true, deep understanding of problems.

It is clear that greater understanding can be achieved by splitting a sentence into individual words, or splitting a problem into its individual sub-units. The relations between those individual sub-units can then be stored, and it is those relationships which generates an understanding of how the world works.

The following diagram appears to capture some of the deep knowledge present in Wolfram Alpha. In the diagram, individual objects are represented as oval shapes, and relationships between objects are represented by arrows – with the type of relationship written on the arrow. For example, it shows that a human has two legs and two arms, but only one head. Whereas it reveals that Zaphod Beeblebrox is an alien who has two heads. Some knowledge of the external environment is also represented on the diagram, showing that humans live on Earth, whereas Beeblebrox lives on a planet near the star of Betelgeuse:

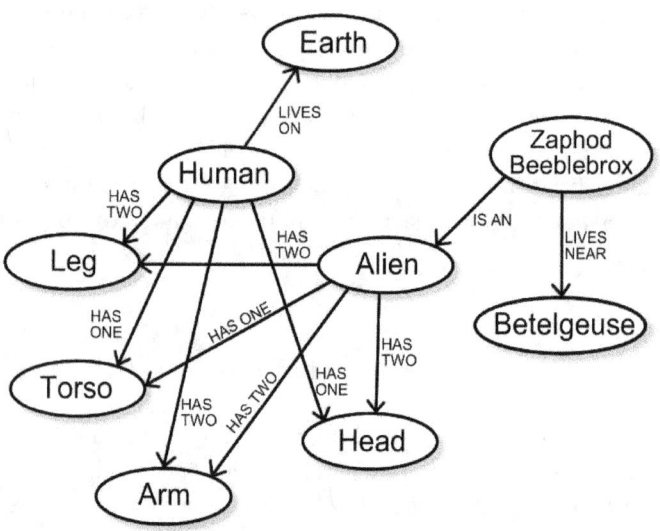

In computer science, this type of diagram is called a *semantic network*. The word "semantics" means "meaning", and a semantic network captures meaning by describing the relationships between objects, and how they affect each other. It would be imagined that knowledge in the brain is stored in a similar way. In psychology, this model of data storage is called an *associative memory*, with different memories being closely associated with other memories. For example, a particular smell may trigger a memory of a meal in a particular restaurant.

The semantic network stored in the brain would form an internal model of the world. The greater the detail of that internal model, the richer and more effective would be our thought processes. For example, we could close our eyes and plan strategies – maybe strategies to hunt animals – and then apply those strategies to the real world. Ideally, we would like our brain's internal model to be as close a match as possible to the real world around us. It is as though we are building a universe inside our brain – or, at least, a simplified copy of the universe in our local vicinity. Indeed, in my first book I explained how the universe is a relative structure, with the functionality of objects having to be described relative to all the other objects in the universe, and meaning arising through relations. As Niels Bohr said: "Isolated material particles are abstractions, their properties being definable and observable only through their interactions with other systems."

We are creating a self-contained universe in our mind!

In his book *The Brain*, David Eagleman considers this internal model in detail. Eagleman stresses that this internal model does not have to be a precise copy of the external world. Indeed, with only a finite number of neurons in the brain, there is no way the internal model could be an atom-for-atom precise copy of the universe. The internal model need only be a fairly accurate representation of the local external world. Eagleman refers to this as a "low-resolution" internal model. As Eagleman explains: "This isn't a failure of the brain. It doesn't try to produce a perfect simulation of the world. Instead, the internal model is a hastily-drawn approximation."

Self-awareness

In Chapter Four, it was explained how surprising new modes of behaviour can "emerge" when a nonlinear network reaches a certain level of complexity. Let us now consider the possibility that self-awareness is one of these emergent behaviours in networks with sufficient complexity, and let us consider if self-awareness holds the key to consciousness.

In a 2017 *New Scientist* article, Michael Graziano, a neuroscientist at Princeton University, stressed that a conscious animal needs "a mental model of its body". Graziano continued: "It's fine for me to say 'arm, go here', but something in my brain needs to have a model of what an arm is, its possible motions, and so on." According to *New Scientist*: "This model is responsible for our conscious awareness of the world, according to Graziano".[28]

Graziano is suggesting that consciousness emerges when the internal model in our brains becomes sufficiently large and detailed that it can contain an accurate model of ourselves, a model which accurately represents our position in the universe and our relations with other objects in that universe. At that crucial moment, we become capable of thinking about ourselves: we become self-aware. And maybe with that self-awareness would come consciousness.

Douglas Hofstadter is currently a professor of artificial intelligence at Indiana University. It was Hofstadter who presented this self-referential theory of consciousness in his classic 1979 book *Gödel, Escher, Bach*. The best description

---

[28] "Why be conscious?", Bob Holmes, *New Scientist*, 13th May 2017.

can be found in the preface of the 20th anniversary edition of the book. In the following brief quote from that preface, Hofstadter explains how consciousness depends on an internal model in the brain which mirrors the external world, and thereby allows self-reference:

> *In short, an "I" comes about – in my view, at least – via a kind of vortex whereby patterns in a brain mirror the brain's mirroring of the world, **and eventually mirror themselves.***

There is a famous geometric structure called a Klein bottle – which is often made of glass – in which the mouth of the bottle is curved around to pass back inside the bottle. Remarkably, this results in the surface of a Klein bottle having only **one side!** If you trace your finger around the "outside" surface you will eventually end up "inside" the surface. There is no distinction between inside and outside: the bottle contains itself.

It seems to me that the "looping effect" of the brain – in which we can think **about** the internal model inside our brain – is reminiscent of the "looping" of the Klein bottle. And with this looping would come self-awareness.

The following diagram shows a glass Klein bottle on the left, and a diagram of the brain on the right containing an internal model of itself. The looping effect – in which we become aware of our own internal model – is shown:

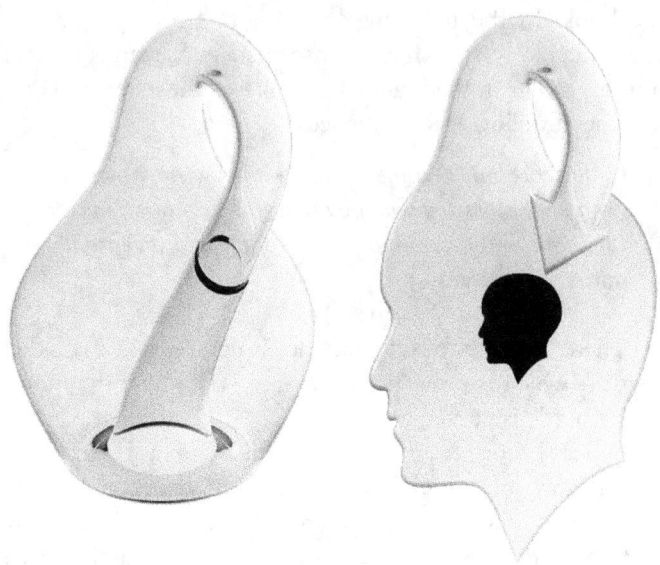

Just as with the Klein bottle, the distinction between the model inside the brain and the person outside the brain becomes blurred. The inside becomes the outside, and the outside becomes the inside. According to Hofstadter, there would then be only one entity: the self-aware, conscious being.

However, I find I cannot totally agree with Hofstadter's theory about self-awareness being the key to consciousness. Remember back to the discussion of the mirror test at the start of this book. It was explained how an animal looking into a mirror would sometimes touch a mark placed on its forehead. This is considered to represent a reliable test of self-awareness. However, it was explained how very few animals have passed the test, and many animals which are surely conscious have failed the test. This seems to indicate that self-awareness is not a completely necessary factor of consciousness.

However, consciousness exists on many levels, dependent on the complexity and richness of the internal model. And Hofstadter makes this point himself:

*When and only when such a loop arises in a brain or in any other substrate, is a **person** – a unique new "I" – brought into being. Moreover, the more self-referentially rich such a loop is, the more conscious is the self to which it gives rise.*

On that basis, it seems the case that self-awareness would be a factor in a **human-level** degree of consciousness (though it might be lacking in lower-level consciousnesses).

Finally, I would like to end this chapter with a short note about Douglas Hofstadter's book *Gödel, Escher, Bach*, which won the Pulitzer Prize and is a masterpiece of popular science writing. In my opinion, *Gödel, Escher, Bach* elevated science writing above the usual unimaginative presentation of dry facts to raise it to the level of an art form (though it does remain a difficult read). As an example, Hofstadter's imaginative style included preceding each chapter with a cryptic dialogue between the fantasy characters created by Lewis Carroll.

Hofstadter's book was remarkable because the imaginative form and structure of the many fantastical dialogues are often just as important as the content of those dialogues. As one example, one of the dialogues is an example of an "acrostic" – a clever section of text which is not what it seems at first glance. The text is arranged so that if you take the first letter of each line of the text, and you note those letters down, you find that those letters form a phrase in themselves (the book contains many self-referential loops like this). I remember the moment I discovered the acrostic in Hofstadter's book – truly an exciting moment and I felt so clever! I remember thinking at the time it would be nice to include an acrostic in my books for readers to find. Someday, one of my books will have a paragraph which includes a similar acrostic, which would be influenced by the genius of Douglas Hofstadter and the great achievement of his *Gödel, Esher, Bach* book. In doing so, that will be my tribute to a piece of science writing which truly inspired me.

8

THE "HARD PROBLEM"

This is the final chapter of this book.

Regular readers of my books will know I like to present some of my own ideas in the final chapter, and this book is no exception.

Up to this point, in considering the necessary conditions for consciousness, this book has only tackled what is known as the "easy" problem of consciousness. In 1995, the Australian philosopher David Chalmers explained how the problem of consciousness is actually two problems:

- **The "easy problem"** – Identifying the requirements necessary for consciousness. Studying the functioning of neurons, identifying which regions of the brain are involved in various conscious tasks, and analysing the structure of the cortex to explain how it can perform intelligent tasks.

- **The "hard problem"** – Explaining how the feeling of consciousness can arise from the processing of information and electrical activity in the brain. The "hard problem" is clearly subjective in nature. How can a feeling arise from the movement of particles, basically.

Chalmers makes the point that the real challenge is to identify the source of the feeling or "experience" of consciousness:

> The really hard problem of consciousness is the problem of **experience.** When we think and perceive, there is a whir of information-processing, but there is also a subjective aspect. This subjective aspect is experience. When we see, for example, we **experience** visual sensations: the felt quality of redness, the experience of dark and light, the quality of depth in a visual field.
>
> But the question of how it is that these systems are subjects of experience is perplexing. Why is it that when our cognitive systems engage in visual and auditory information-processing, we have visual or auditory experience: the quality of deep blue, the sensation of middle C? How can we explain why there is something it is like to entertain a mental image, or to experience an emotion? It is widely agreed that experience arises from a physical basis, but we have no good explanation of why and how it so arises.

I suspect that when people talk about the mystery of consciousness, they are really referring to this "hard problem".

Chalmers believes that conventional neuroscience research is not aimed at uncovering the origin of this experience. Chalmers' paper certainly does not pull its punches:

> The ambiguity of the term "consciousness" is often exploited by both philosophers and scientists writing on the subject. It is common to see a paper on consciousness begin with an invocation of the mystery of consciousness, noting the strange intangibility and ineffability of subjectivity, and worrying that so far we

have no theory of the phenomenon. Here, the topic is clearly the hard problem – the problem of experience. In the second half of the paper, the tone becomes more optimistic, and the author's own theory of consciousness is outlined. Upon examination, this theory turns out to be a theory of one of the more straightforward phenomena. At the close, the author declares that consciousness has turned out to be tractable after all, but the reader is left feeling like the victim of a bait-and-switch. The hard problem remains untouched.

Here is a link to David Chalmers' paper:

http://tinyurl.com/chalmerspaper

In this final chapter, we will be considering the hard problem. There will be no bait-and-switch.

Leibniz's Mill

It was explained in Chapter Two how neurons communicate via electrochemical pulses down their axons. So, at any given moment in time, there will be a pattern of spatially-separated electric pulses in your brain – maybe a considerable distance apart (the pattern of these electric charges is detected in an EEG scan in which electrodes are applied to the scalp). How, then, do I feel like a single conscious entity, rather than feeling like I am just one of those isolated pulses? How can a pattern of isolated electrical pulses produce a single consciousness?

The following image shows the mystery. At any point in time, inside your brain are a series of isolated electric charges, either the action potentials along the axons of neurons or accumulated charge inside the neurons themselves. But then think how you feel, how your consciousness feels like a connected entity. How on Earth can that amazing feeling of consciousness arise from a series of disconnected electric charges?

The problem was described by Gottfried Leibniz in 1714 in a thought experiment which is now called *Leibniz's Mill*. Leibniz's Mill is described by David Eagleman in his book *The Brain*. Leibniz imagined the mind as a large machine, a processing unit, a factory, a mechanism – which seems fair enough. Leibniz imagined walking around a large mill. You would see cogs and levers moving, but, as David Eagleman says:

> *It would be preposterous to suggest that the mill is thinking or feeling or perceiving. How could a mill fall in love or enjoy a sunset? A mill is just made of pieces and parts. And so it is with the brain, Leibniz asserted. If you could expand the brain to the size of a mill and stroll around inside it, you would only see pieces and parts.*
>
> *When we look inside the brain, we see neurons, synapses, chemical transmitters, electrical activity. We see billions of active, chattering cells. Where are you? Where are your thoughts? Your emotions? To Leibniz, the mind seemed inexplicable by mechanical causes.*

So, just are there are disconnected pieces of machinery operating in a mill, so there seems to be disconnected electric charges in the brain. What brings it all together to produce consciousness?

I have seen some ideas written in pseudo-scientific books about a vague unified "field" which link the charges, the field representing consciousness in some way. However, most physicists would reject those ideas. If an effect cannot be objectively observed, measured, and analysed then they would say there can be no convincing motivation for proposing new physics.

Instead, let us try to find solutions based on our comprehensive knowledge of fundamental physics and technology …

The DRAM analogy

If we assume that consciousness is a result of information processing in the brain, how could that possibly result in a feeling of a unified consciousness? Well, as explained in Chapter One, information is always bound to something physical. In the case of the brain, it has been explained how the information processing between neurons takes the form of the physical movement of charged particles – ions – notably sodium and potassium ions. It is surely safe to conclude that the physical substrate of the information held in the brain is these electrically-charged particles.

So we can conclude that **the physical substrate of consciousness must surely be electric charge.**

OK, that's a good start. Now let us expand on this connection between consciousness and electric charge by using an analogy which was suggested by the physicist Max Tegmark.[29]

Tegmark introduced another parallel between thought processes and electronics. How is information stored in modern electronic devices? A few years ago, information might have been stored magnetically on a hard disk which had to be physically rotated, but there is no such large-scale physical movement in the brain. Instead, modern electronic devices such as iPads have now moved to use purely electronic memory which has no moving parts.

---

[29] *Consciousness as a State of Matter*, Max Tegmark, **http://arxiv.org/abs/1401.1219**

A common type of electronic memory is DRAM (*dynamic random-access memory*). Here is a photograph of some DRAM chips attached to a circuit board which you could insert in your computer in order to increase its memory:

DRAM (pronounced "D-RAM") operates by storing information as electric charge on microscopic *capacitors*. A capacitor is composed of two metal plates separated by an insulating material. When a voltage is applied across the two terminals of the capacitor, electric charge is stored on the plates.

Here is a photograph of some common capacitors:

And here is the schematic symbol for a capacitor which is used in circuit diagrams (you can see the two metal plates):

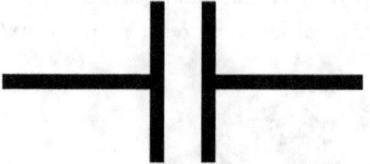

So now let us see how capacitors are used to make DRAM.

In DRAM, each "bit" of information (0 or 1) is stored on a single microscopic capacitor. If the capacitor has electric charge, that represents bit 1, conversely if the capacitor has no electric charge then that represents bit 0. Each capacitor has an associated microscopic transistor, as shown in the following DRAM circuit diagram:

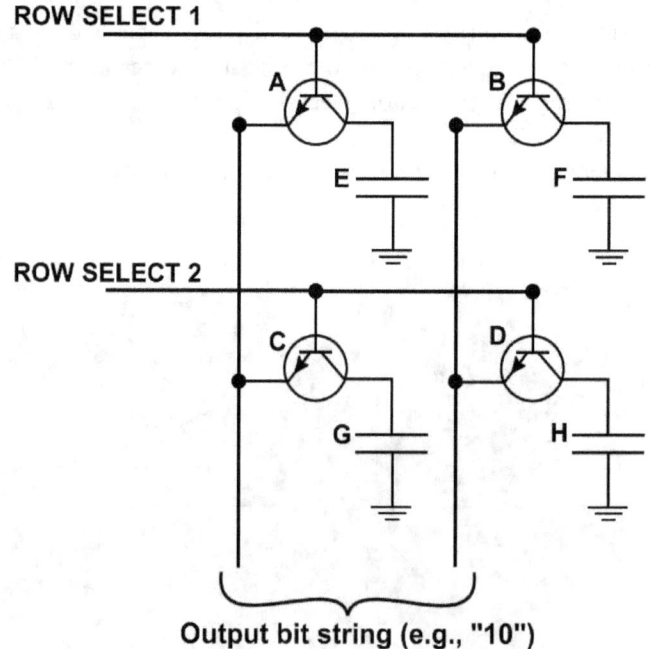

Output bit string (e.g., "10")

You will see that the diagram shows four transistors (labelled A, B, C, and D), and four capacitors (E, F, G, and H). You will recognise the symbol for a transistor from Chapter Three.

Each horizontal row of two capacitors represents a string of two bits of information. As an example, consider the top horizontal row consisting of the two capacitors E and F. If there is electric charge on capacitor E then that would represent bit 1, and if there is no electric charge on capacitor F then that would represent bit 0. So, in that case, the two-bit string stored in the top horizontal row would be "10".

To understand how the data is accessed, let us consider how that data in the top horizontal row might be retrieved. This would be achieved by setting a positive voltage on the top horizontal line "ROW SELECT 1". You will see from the diagram that this voltage is connected to the base leg of transistors A and B. If you remember how a transistor works (as described in Chapter Three), you will see that this applied voltage would have the effect of turning ON transistors A and B. Current is then free to pass between the other two legs of each transistor A and B.

As a result of the two transistors turning ON, you will see from the diagram that the electric charge on the top plate of capacitor E can pass through transistor A and become the first of the output bits (setting the first output bit to bit 1). Also, the electric charge on the top plate of capacitor F can pass through transistor B and become the second of the output bits (setting the second output bit to bit 0).

Therefore, the output string at the bottom of the diagram will be set to "10". And that is how you read the information stored in DRAM. I hope you have enjoyed this little detour into microelectronics.

But you will see that the information stored in DRAM is stored as a large number of spatially-separated electric charges (stored on the capacitors). And, as discussed earlier, **this is just what we see in the human brain.** So let us redraw the earlier diagram of a human brain composed of spatially-separated electric charges, and replace those charges with the charges stored on spatially-separated capacitors. In fact, lets just draw a DRAM brain:

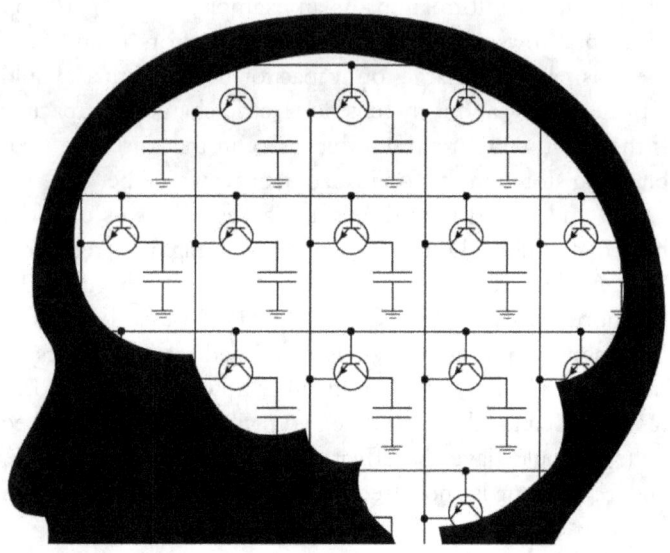

On the previous diagram you can a see a brain composed of an extremely large number of isolated electric charges (stored on capacitors) – just like a real brain.

Now let us return to the previous "Leibniz Mill" thought experiment, and bring it up-to-date. Let us imagine if we could micro-miniaturise Gottfied Leibniz so he was just a few millionths of a metre in height, and let us then implant him inside a DRAM chip. "Mini Leibniz" might walk around the DRAM chip just as he had previously imagined walking around a mill. "Mini Leibniz" would see all the spatially-separated electric charges on capacitors, just as he had previously seen all the isolated pulleys and levers in the mill. And Mini Leibniz might wonder: "How could all these isolated electric charges produce a unified consciousness?", just as he had previously wondered about the mill workings.

But we are fortunate in that we have a greater understanding of technology and DRAM and information than Leibniz possessed. As has just been explained in the DRAM example, we know how to get information out of those "isolated electric charges". We know how computer memory works, and we can extract a string of bits from that DRAM:

```
1011001100110001011011101011011011011101
```

So now we see just what that series of "isolated electric charges" in the brain really represents: it is Claude Shannon's string of bits, **it is raw information.** And, just like a bit string stored in your computer's memory, that raw information might be anything: it might be a movie, or a song, or words in a book. The point is, what seemed like a completely unrelated series of isolated electric charges can actually be a single, unified chunk of information – a single object, in fact – which can be anything.

And that object would be part of a thought in your mind.

This is how Max Tegmark described this analogy in his previously-referenced paper *Consciousness as a State of Matter*:

> *When we view a brain or computer through our physicists' eyes, as myriad moving particles, then what physical properties of the system should be interpreted as logical bits of information? I interpret as a "bit" both the position of certain electrons in my computer's RAM memory (determining whether the micro-capacitor is charged) and the position of certain sodium ions in your brain (determining whether a neuron is firing).*

"It from bit"

So the material of our thoughts is raw information, the physical substrate of which is the pattern of electric charge in our brain. And our internal model of the local universe is composed of this information. We may be under the impression that we live in the physical world, but really we are living in the information-based reality contained within our minds.

David Eagleman describes this principle in his book *The Brain*:

> It feels as though you have direct access to the world through your senses. You can reach out and touch the material of the physical world – like this book or the chair you're sitting on. But this sense of touch is not a direct experience.
>
> Here's the key: the brain has no access to the world outside. Sealed within the dark, silent chamber of your skull, your brain has never directly experienced the external world, and it never will.
>
> Instead, there's only one way that information from out there gets into the brain. Your sensory organs – your eyes, ears, mouth, and skin – act as interpreters. They detect a motley crew of information sources (including photons, air compression waves, molecular concentrations, pressure, texture, temperature) and translate them into the common currency of the brain: electrochemical signals.

The legendary 20th century American physicist John Wheeler expressed a similar idea about the importance of information in creating our world, and the primacy of information over physical reality. Wheeler was born in 1911, and collaborated with Einstein and Bohr. During the war, he worked on atomic bomb development. After the war, Wheeler worked on general relativity and gave one of the most concise and illuminative quotes about that subject:

Space tells matter how to move, and matter tells space how to curve.

Wheeler certainly had a way with words, and was responsible for coining the memorable terms "wormhole" and "black hole".

Here is a photograph of John Wheeler (on the right side of the photograph) taken in 1985, standing by the side of the German physicist Eckehard Mielke. Wheeler has written: "The black hole is a source of enlightenment":

Wheeler remained an active physicist into his 90s. He was never afraid to tackle what he called the "Really Big Questions" and, in 1989, he wrote an influential paper in which he described a principle which he called "It from bit".[30] The idea of "It from bit" is that the physical world (the "It") is derived from the more fundamental world of information (the "bit" – as in Claude Shannon's "bits"). Wheeler explained how our picture of reality is solely derived from the information we can obtain:

> *Every physical quantity, every **it**, derives its ultimate significance from **bits**. That which we call reality arises in the last analysis from the posing of yes-no questions and the registering of equipment-evoked responses; in short, that all things physical are information-theoretic in origin.*

Anton Zeilinger, Director of the Institute for Quantum Optics and Quantum Information, explains his interpretation of John Wheeler's idea:

> *My interpretation is that in order to define reality, one has to take into account the role of information: mainly the fact that whatever we do in science is based on information which we receive by whatever means.*

Though we might not realise it, our personal universe is a universe made of information.

---

[30] *It From Bit*, John Wheeler,
http://tinyurl.com/wheelerinformation

Beyond the physical

In his paper, David Chalmers presented what I consider to be a great insight. Chalmers presented the highly-convincing argument that merely altering the arrangement of physical material is never going to provide us with an explanation of the feeling of consciousness. All we will ever be doing is moving physical material around into various different configurations:

> *For any physical process we specify there will be an unanswered question: why should this process give rise to experience? The structure and dynamics of physical processes yield only more structure and dynamics, so structures and functions are all we can expect these processes to explain.*

To my mind, this makes a great deal of sense. Why should merely changing the position of physical objects (particles, for example) cause the emergence of an apparently completely unconnected phenomenon: the feeling of consciousness?

This is basically the same argument as in the Leibniz Mill thought experiment. If you remember, the brain was compared to mill machinery. But if you walked around that mill you would just see the movement of cogs and levers, "pieces and parts" as David Eagleman described. Where, then, does the feeling of consciousness come from?

However, in the previous DRAM analogy, we have seen that the arrangement and movement of physical material can represent something which is not physical: information.

According to Chalmers, we need to find an "extra ingredient" in the explanation. I believe that "extra ingredient" is information.

The reason why I believe that is the case is because – when we examine Chalmers' argument – it is clear that we **need** to go **beyond the physical** to find a solution to the hard problem.

This emphasis on information represents something of a departure from physics, because physics is about physical things (hence the name), and raw information is not physical (you cannot hold or touch raw information). Things can certainly be made of information, but those things would be movies, or digital images, or music, for example. Those things exist as independent entities – but they are not physical things. I am therefore suggesting that consciousness is not a physical thing. Information always requires a physical substrate, but it seems to exist in a realm **above** the physical world. Take away the physical substrate and consciousness disappears – but consciousness is not physical.

At this point, allow me to introduce a brand-new scientific term: the *thing*. I am going to define a "thing" as the most general type of object that can possibly exist. To be precise, a "thing" is going to be defined as "anything that can be made from anything".

OK, that sounds extremely vague. How can such a vague term possibly be of any use? Well, perhaps surprisingly, it appears that a "thing" can only fall into one of two possible groups. Firstly, a thing can be **physical**, in which case it would be composed of particles contained in the Standard Model. That would be fermionic matter (matter made of atoms) such as cars and trees. But it would also include objects made of immaterial bosons, such as rays of light and radio waves (items made of photons – particles described within the Standard Model). Clearly, that is a huge group.

The other grouping is equally huge. This would be the objects made purely out of information. This would include a range of "things" from digital media files such as digital movies and music, through to a thought in your mind (made out of information represented by the arrangement of electrical charge in your brain).

Here is a diagram showing how every "thing" has to fall into one of the two categories:

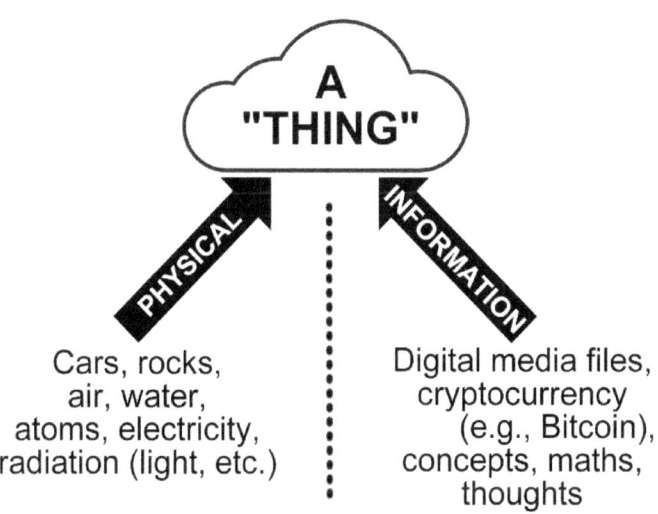

You will see that I have included cryptocurrency – such as Bitcoin – in the second grouping. This is digital money which has **no physical form.** Conventional currencies always have a physical form, e.g., coins and banknotes. However, Bitcoin only exists as information, stored and transmitted in electronic format. Bitcoin is therefore clearly not a physical "thing". As I said earlier, you cannot hold or touch raw information, so you cannot hold or touch a Bitcoin. However, Bitcoin might be considered to be just as real as the coins and notes in your wallet.

You will also see that I have included "concepts" in the second grouping. Basically, if you can represent a "thing" in its entirety by thinking about it or writing it down, then it is made of information and it should be in the second group.

Into which group should we place a colour? For example, into which group should we place "green"? That is an interesting question. Firstly, "green" could refer to the particular wavelength of light which we associate with the colour (approx. 495-570 nm), in which case – as light is listed on the diagram as being a physical "thing" – we should place it in the first group. However, the feeling which we experience when we look at something which is coloured green – the "feeling" of green – is clearly a thought and should therefore be placed in the second group. So a colour has a dual interpretation and could therefore be placed in either group depending on that interpretation.

So there is clearly a bridge between the two groups: physical colour can be converted into the thought of colour in your mind, or can be related to that feeling of colour. So there is not an unbridgeable gap: the mind/body gap can be crossed. Information is associated with all physical "things". Far from there being an unbridgeable gap, there is actually an unbreakable link.

I can think of no "thing" which does not fall into one of these two categories. If you can think of any "thing" which is outside of these two categories, my email address is in the front of the book – please let me know!

I think the fact that absolutely every "thing" in existence can be placed in one of only two categories must have some profound significance. Might this be interpreted as Descartes' "dualism"?

As stated earlier, on the basis of Chalmers' argument, in order to solve the hard problem we **need** to go **beyond the physical.** We need to move to the other grouping of "things". Referring to the previous diagram, the question as to whether or not a particular material could ever become conscious then reduces to a very simple black-or-white question: **on which side of the line does it lie?** If it lies on the right-hand side of the line, then the material has the potential to be conscious.

I believe this "What side of the line?" principle provides us with a huge clue about how to solve the hard problem of consciousness. **I would suggest that we could define the "feeling" of consciousness as the feeling you get when you look at a physical object (a car, rock, etc.) and know that you are made from different stuff.** In other words, it is the feeling you get from knowing you are on the right-hand side of the line in the previous diagram, as opposed to being on the left-hand side of the line in the previous diagram. It is the feeling of being "made differently" that defines consciousness, the feeling that your consciousness is made of information.

In a nutshell, **consciousness is the feeling that your mind is not a physical thing.** You feel different because you **are** different: you are a fundamentally different type of "thing".

And how would you actually feel if you were on the right-hand side of the previous diagram? Well, in Chapter One it was explained how a consciousness based on information would feel substrate-independent (you would not feel as though you were made of a "rice pudding brain"), feeling independent of the physical world, with a mind which felt impervious to physical damage and the ravages of time. This is clearly just how your consciousness feels. This is a view shared by the physicist Max Tegmark in his previously-referenced paper:

> *I have long contended that consciousness is the way information **feels** when processed in certain complex ways.*

So, to an extent, it is possible to explain the "feeling" of consciousness via this information-based model. It seems that science can provide insights even into subjective matters.

So, yes, I am essentially suggesting that your consciousness is like Bitcoin: made of information with no physical form, and completely substrate-independent:

I would like to leave you with one final thought.

We live in a world in which science has stripped Nature down to her smallest constituent parts, and eliminated almost all mysteries about the natural world – from the universe to the atom. So perhaps we should be rather proud of the fact that one of the few remaining mysteries of Nature lies at the heart of each and every one of us.

PICTURE CREDITS

All photographs are public domain unless otherwise stated.

Photograph of baboon performing the mirror test is by Moshe Blank and is provided by Wikimedia Commons.

Photograph of Claude Shannon is by Konrad Jacobs and is provided by Wikimedia Commons.

Photograph of the Boltzmann entropy formula is by Daderot and is provided by Wikimedia Commons.

Photograph of the resistor is by Nunikasi and is provided by Wikimedia Commons.

Termite mound photograph is by RayNorris and is provided by Wikimedia Commons.

Futuristic cityscape image is a composite of a public domain image and my own work.

Photograph of the DER2 android is by Gnsin and is provided by Wikimedia Commons.

Photograph of the capacitors is by Eric Schrader and is provided by Wikimedia Commons.

ACKNOWLEDGEMENTS

Thanks to Alexander Bystritsky for ideas and discussion.

HIDDEN
INPLAINSIGHT10

How to program a
quantum computer

AVAILABLE NOW